The Gospel of Mark

William MacDonald

EMMAUS
WORLDWIDE

Developed as a study course by Emmaus Correspondence School, founded in 1942.

The Gospel of Mark
William MacDonald
Published by:
  Emmaus Worldwide
  PO Box 1028
  Dubuque, IA 52004-1028
  phone: (563) 585-2070
  email: info@emmausworldwide.org
  website: EmmausWorldwide.org

First Printed 1970 (AK '70), 1 Unit
Revised 2005 (AK '05), 1 Unit
Reprinted 2007 (AK '05), 1 Unit
Revised 2011 (AK '11), 2 Units
Reprinted 2013 (AK '11), 2 Units
Revised 2020 (AK '20), 2 Units

ISBN 978-1-59387-481-0

Code: ER11MK

Printed in the United States of America

Course Overview

The Gospel of Mark follows Jesus moving from one event to another, submitting to the Father's will at every turn. We see Him healing the sick, opening the eyes of the blind, forgiving sinners and eating with the outcasts of society. Mark tightly weaves these stories together culminating with the ultimate sacrifice of Jesus's own life, showing us that Jesus "came not to be served but to serve, and give his life as a ransom for many" (Mark 10:45). As you work your way through this course, we hope it will challenge you to follow Jesus's example in how He lived among people---as a humble Servant.

Lessons You Will Study

Student Instructions

This Emmaus course is designed to help you *know God* through a *better understanding of the Bible* and *how it applies to your life*. However, this course can never take the place of the Bible itself. The Bible is inexhaustible, and no course could give the full meaning of its truth. If studying this course is the end goal, it will become an obstacle to your growth; if it is used to inspire and equip you for your own personal study of the Bible, then it will achieve its goal. As you study the Bible using this course, prayerfully ask God to reveal His truth to you in a powerful way.

Course Sections

This course has three parts: the *lessons*, the *exams*, and an *exam sheet*.

The Lessons

Each lesson is written to help explain truths from the Bible. Read each lesson through at least twice—once to get a general idea of its content, and then again, slowly, looking up any Bible references given. You should always have your Bible opened to the verses or passage being studied. It is important that you read the Bible passages referenced, as some questions in the exams may be based on the Bible text.

To look up a Bible verse, keep in mind passages in the Bible are listed by book, chapter, and verse. For instance, 2 Peter 1:21 refers to the second book of Peter, chapter 1, and verse 21. At the beginning of every Bible, there is a table of contents which lists the names of the books of the Bible, and tells the page number on which each book begins. For practice, look up 2 Peter in the table of contents and turn to the page number listed; then find the chapter and verse.

The Exams

At the end of each lesson, there is an exam to assess your knowledge of the course material and the Bible passages. The exams contain multiple choice and/or True/False (T/F) questions. After you have studied a lesson, complete the exam for that lesson by recording your answers on the exam sheet at the end of the course. If you have difficulty answering the questions, re-read the lesson or use the Bible as a reference.

Please note, it is best not to answer the questions based on what you *think* or have *always believed*. The questions are designed to find out if you understand the material in the course and the Bible.

What Do You Say?

In addition to the multiple choice section, each exam also contains a *What Do You Say?* question. These questions are designed for your personal reflection and to help you express your ideas and feelings as you process the lesson's content. Your responses should not be written on the attached exam sheet, but rather, directly written in the space provided at the end of each exam.

The Exam Sheet

Use the exam sheet included at the end of the course to complete the exams. When you have determined the right answer to a question on an exam, fill in the corresponding letter on the exam sheet. If you have an Emmaus Connector and he/she provides a different exam sheet, please use that one instead.

Write It Out!

The exam sheet also contains *Write It Out!* questions. These questions are designed to help you apply the course's content to your daily life.

Continued on next page

Write It Out! (continued)

The *Write It Out!* questions will address your *head* (thinking), *heart* (feeling), and *hands* (doing).

1. **Head.** The first *Write It Out!* question will be geared toward your head and asks you to respond to a critical question concerning the course in its entirety.

2. **Heart.** The next question asks how the course affects your perspective of or feelings toward God, yourself, or others.

3. **Hands.** The final question asks you what action you need to take in response to what you have learned.

Write It Out! questions will be reviewed and responded to by an Emmaus Connector, an official partner of Emmaus Worldwide who is trained to encourage and guide you with their response.

Prayer Requests or Questions?

You may include your personal questions on the exam sheet to help us get to know you and your needs better. Please let us know specific questions you may have about the Bible, God, or other spiritual matters. You may also include a personal prayer request and we will pray for you.

Submitting the Exam Sheet

When you have answered all the exam questions on the exam sheet, check them carefully. Slowly tear out the exam sheet along the perforated edge near the course spine. Please note, do not tear out the exams from the lessons; submit the *exam sheet only*.

Fill in your contact information and submit your completed exam sheet to your Emmaus Connector or the organization from which you received it (several options for submission are shown at right).

OPTION 1: Send to your Emmaus Connector

If you know your Emmaus Connector, give them your completed exam sheet or mail it to the address listed here (if blank, go to option 2).

OPTION 2: Send to Emmaus Worldwide's head office

If no address is listed above, or you do not know if you have an Emmaus Connector and are unsure of where to send your exam sheet, you can:

MAIL the exam sheet to

Emmaus Worldwide
PO Box 1028
Dubuque, IA 52004-1028

OR

EMAIL a scan or photo

of both sides of the exam sheet to this email address:

Exams@EmmausWorldwide.com

Receiving Your Results

You will receive your graded exam sheet back (through the same method it was submitted, either mail or email), including your final grade and personal response from an Emmaus Connector.

After finishing this course, you will be awarded a certificate of completion, which serves as official record that you studied the contents of this course. A transcript of courses you complete will be stored by Emmaus Worldwide or your Emmaus Connector.

Introduction

In this Gospel, we have the wonderful story of God's Perfect Servant, our Lord Jesus Christ. It is the story of One who laid aside the outward display of His glory in heaven and assumed the form of a servant on earth (Philippians 2:7). It is the story of One who "went about doing good and healing all who were oppressed by the devil" (Acts 10:38). It is the matchless story of One who "came not to be served but to serve, and to give His life as a ransom for many" (Mark 10:45).

If we remember that this Perfect Servant was none other than God the Son, and that He willingly girded Himself with the apron of a bond-slave, becoming a servant of men, the Gospel will glow with constant splendor.

Here we see the Incarnate Son of God living as a dependent Man on earth. Everything He did was in perfect obedience to His Father's will, and all the mighty works He performed were done in the power of the Holy Spirit.

The book was written by John Mark, a servant of the Lord who is mentioned in Acts 15:37-38 and 2 Timothy 4:11.

It is generally agreed that Mark was influenced greatly by Peter in writing this Gospel. The early church fathers spoke of Peter as one who supplied Mark with much of the eyewitness material.

Mark's style is rapid, energetic, and concise. He emphasizes the deeds of the Lord more than His words. This is seen by the fact that he records nineteen miracles, but only four parables.

As we study the Gospel, we will discover answers to these questions:

1. What does it say?
2. What does it mean?
3. What lessons are there in it for me?

The Gospel of Mark present Jesus as the perfect example of servanthood. The heart of the Gospel can be seen in Mark 10:45: "For even the Son of Man came not to be served but to serve, and to give His life as a ransom for many."

The Servant Begins His Work

Mark 1

Herald of the Good News (1:1-8)

Mark's theme is the good news about Jesus Christ, the Son of God. Because his purpose is to emphasize the servant role of the Lord Jesus, he does not begin with a genealogy. Rather he begins with the public ministry of the Savior. This was announced by John the Baptist, the herald of the good news.

Both Malachi and Isaiah had predicted that a messenger would precede the Messiah, calling on the people to be morally and spiritually prepared for His coming (Malachi 3:1; Isaiah 40:3). These prophecies were fulfilled in John the Baptist. He was "the messenger" and "voice cries: in the wilderness."

His message was that the people should repent in order to receive the forgiveness of sins. Repentance means they should confess and forsake their sins.

> **Mark's theme is the good news about Jesus Christ.**

When his hearers did repent, John baptized them as an outward expression of their about-face. Baptism separated them publicly from the mass of the nation of Israel who had forsaken the Lord. It united them with a remnant who were ready to receive the Christ.

It would seem from verse 5 that the response to John's preaching was universal. This was not the case. There may have been an initial burst of enthusiasm, with multitudes surging out to the desert to hear the fiery preacher, but the majority of the people did not genuinely confess and forsake their sins. This will be seen as the narrative advances.

What kind of man was John? Today he would be called a fanatic and an ascetic. His home was the desert. His clothing was the coarsest and the simplest. His food was sufficient to maintain life and strength, but was scarcely the diet of luxury. He was a man who subordinated all these things to the glorious task of making Christ known. Perhaps he could have been rich, but he chose to be poor. He thus became a fitting herald of Him who had nowhere to lay His head. We learn here that simplicity should characterize all who are servants of the Lord.

His message was the superiority of the Lord Jesus. He said that Jesus was greater in power, greater in personal excellence, and greater in ministry. John did not consider himself worthy to untie the Savior's shoelaces—the menial duty of a slave. Spirit-filled preaching always exalts the Lord Jesus and dethrones self.

John's baptism was in water. It was an external symbol, but produced no change in a person's life. The Lord Jesus would baptize with the Holy Spirit; this baptism would produce a great inflow of the Spirit's power (Acts 1:8). Also it would incorporate all believers into the church, the body of Christ (1 Corinthians 12:13).

Jesus Baptized in the Jordan (1:9-11)

The Gospels do not record much of Jesus' life before the age of thirty. After these thirty "silent years" in Nazareth, the Lord Jesus was ready to enter upon His public ministry. First He traveled the sixty odd miles from Nazareth to the Jordan near Jericho. There He was baptized by John. In His case, of course, there was no repentance because there were no sins to confess. Baptism for the Lord was a symbolic action picturing His eventual baptism into death at Calvary and His rising from the dead. Thus at the very outset of His public ministry, there was this vivid foreshadow of a cross and an empty tomb.

As soon as He came out of the water, the heavens opened and the Holy Spirit descended on Him as a dove. The voice of God the Father was heard, acknowledging Jesus as His beloved Son.

The power of the Holy Spirit is indispensable.

The Holy Spirit came upon Him, anointing Him for service and equipping Him with power. It was a special ministry of the Spirit, preparing Him for the three years of service that lay ahead.

The power of the Holy Spirit is indispensable. A person may be educated, talented and fluent, yet without the inner work of the Holy Spirit, their service is lifeless and ineffective. The question is basic, "Is the Holy Spirit empowering my service to the Lord?"

Into the Wilderness (1:12-13)

After His baptism, Jesus was tempted by Satan in the wilderness for forty days. It was the Spirit of God who led Him to this rendezvous to prove that He would not sin. His victory over temptation makes him a sympathetic and effective priest for us (Hebrews 4:14-16).

Why does Mark say that He was with the wild animals? Were these animals energized by Satan to seek to destroy the Lord? Or were they docile in the presence of their Creator? We can only ask the questions.

It was at the end of the forty days that the angels ministered to Him (Matthew 4:11). During the temptation, He ate nothing (Luke 4:2).

Testings are inevitable for the believer. The closer one follows the Lord, the more intense they will be. Satan does not waste his gunpowder on nominal Christians. But he opens his big guns on those who are winning territory in the spiritual warfare.

It is not a sin to be tempted. The sin lies in yielding to temptation. In our own strength we are not able to resist. But the indwelling Holy Spirit is the believer's power to subdue dark passions.

Beginning of the Galilean Ministry (1:14-15)

Mark skips over the Judean ministry of the Lord (see John 1:1-4:54) and begins with the great Galilean ministry. This extends from Mark 1:14 to 9:50 and covers a period of one year and nine months. Then he deals briefly with the latter part of the Perean ministry (10:1-10:45) before moving on to the last week in Jerusalem.

Jesus went to Galilee, preaching the good news about God. His specific message was that:

- The time was fulfilled. According to the prophetic time-table, a date had been fixed for the public appearing of the King. That date had now arrived.
- The kingdom of God was at hand. In other words, the King was present and was making a bona fide offer of the kingdom to the nation of Israel. The kingdom was at hand in the sense that the King had appeared on the scene.

• People were called on to repent and to believe the Gospel. In order to be eligible to enter the kingdom, they had to do an about-face with regard to sin. Also they had to believe the good news concerning the Lord Jesus.

Fishers of Men (1:16-20)

As He walked along the shore of the Sea of Galilee, Jesus saw Simon and Andrew fishing. He had met them before; in fact, they had become disciples of His at the outset of His ministry (John 1:40-41). Now He called them to be with Him, promising to make them fishers of men. Immediately they gave up their lucrative fishing business to follow Him. Their obedience was prompt, sacrificial and complete.

Fishing is an art, and so is soul-winning.

• It requires patience. Oftentimes there are lonely hours of waiting.
• It requires skill in the use of bait, lures or nets.
• It requires discernment and common sense in going where the fish are running.
• It requires persistence. A good fisherman is not easily discouraged.
• It requires quietness. The best policy is to avoid disturbances and to keep self in the background.

We became fishers of men by following Christ. The more like Him we are, the more successful we will be in winning others to Him. Our responsibility is to follow Him; He will take care of the rest.

A little further on, the Lord Jesus met James and John, the sons of Zebedee, as they were mending their nets. As soon as He called them, they said goodbye to their father and went after the Lord.

Christ still calls people to forsake all and follow Him (Luke 14:33). Neither possessions nor parents must be allowed to hinder obedience.

Miracles in the Synagogue (1:21-28)

In verses 21-38, we have a typical twenty-four hour day in the life of the Lord. Miracle followed miracle as the Great Physician healed those who were demon-possessed and diseased.

The healing miracles of Christ illustrate how the Savior of the world liberates people from the dread results of sin. We may illustrate this as follows:

Miracle:	Deliverance from:
1. Healing of a man with unclean spirit (1:23-26).	1. The uncleanness of sin.
2. Healing of Simon's mother-in-law (1:29-31).	2. The feverishness and restlessness of sin.
3. Healing of the leper (1:40-45).	3. The loathsomeness of sin.
4. Healing of the paralytic (2:1-12).	4. The helplessness caused by sin.
5. Healing of man with withered hand (3:1-5).	5. The uselessness caused by sin.
6. Deliverance of the demoniac (5:1-20).	6. The misery, violence and terror of sin.
7. The woman with the discharge of blood (5:25-34).	7. Sin's power to sap life's vitality.
8. The raising of Jairus' daughter (5:21-24; 35-43).	8. Spiritual death caused by sin.
9. Healing of Syrophenician's daughter (7:24-30).	9. The thralldom of sin and Satan.
10. Healing of deaf man with speech impediment (7:31-37).	10. Inability to hear God's Word and to speak of spiritual things.
11. Healing of blind man (8:22-26).	11. Blindness to the light of the Gospel.
12. Healing of demoniac boy (9:14-29).	12. The cruelty of Satan's dominion.
13. Healing of blind Bartimaeus (10:46-52)	13. The blind and beggarly state to which sin reduces.

Though the preacher of the Gospel is not called upon to perform these acts of physical healing today, he is constantly called upon to deal with their spiritual counterparts. Are these not the greater miracles which the Lord Jesus mentioned in John 14:12: "Truly, truly, I say to you, whoever believes in me will also do the works that I do; and greater works than these will he do, because I am going to the Father."

But now let us get back to the synagogue in Capernaum. Jesus had entered on the Sabbath and begun to teach. The people realized that here was no ordinary teacher. There was an undeniable power connected with His words. He was not like the scribes who droned on mechanically. His sentences were arrows from the Almighty. His lessons were arresting, convicting, challenging. The scribes peddled a second-hand religion. But there was no unreality in the teaching of the Lord Jesus. He had the right to say what He did, because He lived what He taught.

Everyone who teaches the Word of God should speak with authority. If we cannot speak with assurance, we ought not to speak at all. The psalmist said, "I believed, therefore I spoke" (Psalm 116:10 NKJV) and Paul echoed the words in 2 Corinthians 4:13. Their message was born of deep conviction.

> **Jesus had the right to say what He did, because He lived what He taught.**

In the synagogue, there was a demon-possessed man. The particular type of demon that inhabited him is described as an unclean spirit. This probably means that the spirit manifested its presence in the man by making him physically or morally unclean.

Let no one confuse demon possession with various forms of insanity. The two are separate and distinct. Demon-possession means that a person is actually indwelt and controlled by an evil spirit. The person is often enabled to perform supernatural feats and they often become violent or blasphemous when confronted with the Person and work of the Lord Jesus Christ.

Notice that in verse 24, the evil spirit recognized Jesus and spoke of Him as Jesus of Nazareth and the Holy One of God.

Notice too the change of pronouns from plural to singular: "What have you to do with *us*? Have you come to destroy *us*? *I* know who you are." At first the demon speaks as joined to the man, then he speaks for himself alone.

Jesus would not accept the witness of a demon, even if it were true. So He told the evil spirit to be silent, and then commanded him to come out of the man. It must have been a weird experience to see the convulsed man and to hear the loud, eerie cry of the demon as he left his victim.

The miracle caused amazement among the people. This was new and startling to them—that with a mere command a Man could drive out a demon. Did this mark the beginning of a new school of religious teaching, they wondered? Quickly the news of the miracle spread throughout Galilee.

Before leaving this portion, there are one or two comments that should be made:

- It seems that the first advent of Christ aroused a great outburst of demonic activity on the earth.

- Christ's power over these evil spirits is a foreshadow of His eventual triumph over Satan and all his agents.
- Wherever God works, Satan opposes. All who set out to serve the Lord can expect to be opposed every step of the way. "For we do not wrestle against flesh and blood, but against the rulers, against the authorities, against the cosmic powers over this present darkness, against the spiritual forces of evil in the heavenly places" (Ephesians 6:12).

Simon's Mother-in-law Healed (1:29-31)

"Immediately" is one of the characteristic words of this Gospel. It occurs nine times in this chapter alone. This and other similar words are specially suitable for the Gospel that stresses the servant character of the Lord Jesus.

From the synagogue, our Lord went to Simon's house. As soon as He arrived, he learned that Simon's mother-in-law was ill with a fever. They didn't waste any time in bringing a case of genuine need to the Physician's attention.

Without saying a word, Jesus took her by the hand and helped her to her feet. She was instantly cured.

Ordinarily a fever leaves a person in an extremely weakened condition. In this case, the Lord not only cured the fever but gave immediate strength to serve, "she began to serve them." "There is a whole cluster of suggestions here. Every sick person who is restored, whether in an ordinary or extraordinary way, should hasten to consecrate to the service of God the life that is given back. Surely it was spared for a purpose, and we shall be disloyal to God if we do not thus devote it. A great many persons are always sighing for opportunities to minister to Christ, imagining some fine and splendid service that they would like to render. Meantime they let slip past their hands the very things in which Christ wants them to serve Him. True ministry to Christ is doing first of all and well one's daily duties." (J. R. Miller, *Come Ye Apart* (New York: Thomas Crowell & Co., 1887). Daily Reading for March 28.)

In studying the healing miracles, it is noticeable that the Savior's procedure is different in every case. This reminds us that no two cases of conversion are exactly alike. Everyone must be dealt with on an individual basis.

The fact that Peter had a mother-in-law shows that God did not require His servants to be single. Enforced celibacy is a human tradition of that finds no support in the Word of God. Besides, it subjects men and women to terrible temptation.

Healing at Sunset (1:32-34)

News of the Savior's presence had spread during the day. As long as it was the Sabbath, the people dared not bring the needy to Him. But as soon as the sun had set and the Sabbath had ended, there was a rush to the door of Peter's house. There the diseased and the demon-possessed experienced the power that delivers from every phase and form of sin.

Twice in these verses, demon possession is distinguished from sickness. Those who have witnessed the former know that it is not the same as insanity.

First Circuit of Galilee (1:35-39)

Jesus rose long before daybreak, went out to a place where He would be free from distraction and spent time in prayer. The Servant of Jehovah opened His ear each morning to receive instructions for the day from God the Father (see Isaiah 50:4-5).

If the Lord Jesus felt the need of this early morning quiet time, how much more should we! Notice too that He prayed when it cost Him something; He rose up and went out a great while before day. Prayer should not be a matter of personal convenience but of self-discipline and sacrifice. Does this explain why so much of our service is ineffective?

> In the morning, first of all,
> Savior, may I hear Thy call;
> Make me ready to obey
> Thy commands throughout the day.

By the time that Simon and the others got up, the crowd was gathered outside the house again. The disciples went out to tell the Lord of the rising popular sentiment. Surprisingly enough, He did not go back to the city but took the disciples into the surrounding towns, explaining that He must preach there also.

Why did He not return to Capernaum?

- First of all, He had just been in prayer and had learned what God wanted Him to do that day.

- Secondly, He realized that the popular movement in Capernaum was shallow. The Savior was never attracted by large crowds. He looked below the surface to see what was in their hearts.
- He knew the peril of popularity and taught the disciples by His example to beware when all spoke well of them.

> **Jesus prayed when it cost Him something.**

- He consistently avoided any superficial, emotional demonstration that would have put the crown before the cross.
- His great emphasis was on preaching the Word (vv. 38-39). The miracles of healing, while intended to relieve human misery, were also designed to gain attention for the preaching.

Thus to the synagogues of Galilee, Jesus went preaching and casting out demons. He combined preaching and practicing, saying and doing.

The Prayer God Answers (1:40-45)

The account of the leper gives us an instructive example of the prayer that God answers.

- It was earnest and desperate—"imploring Him."
- It was reverent—"kneeling said to Him."
- It was humble and submissive—"If you will."
- It was believing—"You can."
- It acknowledged need—"make me clean."
- It was specific—not "bless me" but "make me clean."
- It was personal—"make ME clean."
- It was brief—five words in the original.

Notice what happened!

- Jesus was moved with compassion (NKJV). Let us never read these words without a sense of exultation and gratitude.
- He stretched forth His hand. Think of it! The hand of God stretched forth in answer to humble, believing prayer.
- He touched him. Under the law, a person became ceremonially unclean when they touched a leper. Also, of course, there was the danger of contracting the disease. But the Holy Son of Man identified Himself with the miseries of mankind, dispelling the ravages of sin without being tainted by them.

- He said, "I will." He is more willing to heal than we are to be healed. Then "Be clean." In an instant the skin of the leper was smooth and clear.
- He forbade publicizing the miracle until first the man had appeared before the priest and had made the required offering (Leviticus 13:49ff; 14:2ff). This was a test, first of all, of the man's obedience. Would he do as he was told? He did not; he publicized his case, and as a result, he hindered the work of the Lord (v. 45). It was also a test of the priest's discernment. Would he perceive that the long-awaited Messiah had come, performing wonderful miracles of healing? If he was typical of the nation of Israel, he did not.

Again we find that Jesus withdrew from the crowds and ministered in desert places. He did not measure success by numbers.

LESSON 1 EXAM

Use the exam sheet at the back of the course to complete your exam.

1. **The baptism of Jesus**
    A. symbolized repentance towards God.
    B. was the first New Testament example of Christian baptism.
    C. symbolized His eventual baptism into death and subsequent resurrection.
    D. publically incorporated Him into the Jewish nation.

2. **The descent of the Holy Spirit upon the Lord Jesus at His baptism signified**
    A. that the dispensation of Law was over.
    B. the filling of the Lord Jesus with the Spirit.
    C. the anointing of Jesus with power for service.
    D. that Jesus was "the Lord from heaven."

3. **The purpose of the Lord's temptation in the wilderness was to**
    A. see if He would sin.
    B. prove He could not sin.
    C. defeat the Devil once and for all.
    D. demonstrate the Lord's power over evil beasts.

4. **Mark 1:14-9:50 covers a period of one year and nine months and deals with**
    A. the Lord's Judean ministry.
    B. the Lord's Perean ministry.
    C. the Lord's Galilean ministry.
    D. the Lord's ministry in Galilee and in Perea.

5. **Which of the following was not a part of the Gospel as preached by the Lord Jesus?**
    A. The kingdom of God was at hand.
    B. Believers in the Lord Jesus must be baptized.
    C. Men must repent and believe the Gospel.
    D. The time had come for God's King to be publicly revealed.

6. **Demon possession**
   A. is restricted especially to the days when Jesus was on earth.
   B. ought, strictly speaking, to be classified as insanity.
   C. often makes a person violent or blasphemous when confronted with the Person and work of Christ.
   D. never manifests itself in verbal acknowledgment of who Jesus really is.

7. **A characteristic word of Mark's Gospel is**
   A. "afterwards".   C. "whosoever".
   B. "nevertheless".   D. "immediately".

8. **The healing miracles of Jesus and cases of conversion have this in common—**
   A. both take place without the consent of the individual.
   B. no two cases are alike.
   C. the event always takes place at a public gathering.
   D. both are psychologically explained.

9. **One reason why Jesus left Capernaum was because**
   A. popular sentiment was against Him.
   B. it was impossible to do many mighty works because of the unbelief of the people
   C. the people kept Him too busy to pray.
   D. He knew the peril of popularity and wished to warn His disciples against it.

10. **The leper who came to Jesus was sent by the Lord to the priest**
   A. in order to be healed by the priest.
   B. as a testimony to the priest.
   C. because the Lord wished to demonstrate the impossibility of a man being saved by mere religion.
   D. because the Lord did not wish to offend the religious leaders.

What Do You Say?

What can we learn about prayer from the prayer of the leper?

Early Opposition

Mark 2, 3

Paralysis Exchanged for Power (2:1-12)

It wasn't long after the Lord returned to Capernaum that the crowd began to gather around the house where He was. The word had spread quickly, and the people were anxious to see the Miracle Worker at work. Whenever God moves in power, people are attracted.

The Savior faithfully preached to them as they clustered around the door. At the rear of the crowd was a paralyzed man, carried by four others on an improvised stretcher. No reason is given for his paralysis. The crowd was a hindrance to his getting near to the Lord Jesus.

There usually are hindrances in bringing others to Jesus. But faith is ingenious. The four carriers climbed up the outside stairs to the roof, uncovered a portion of the roof, and lowered the paralytic to the ground floor—perhaps to a courtyard in the middle.

> **Whenever God moves in power, people are attracted.**

That brought him near the Son of God. It is good to have friends like that. Someone has nicknamed them Sympathy, Cooperation, Originality, and Persistence. We should all strive to be a friend who displays these qualities.

Jesus was impressed by their faith—the faith of all five of them, no doubt. He turned to the helpless man and said, "Son, your sins are forgiven."

Now this seemed to be a strange thing to say. It was a question of paralysis, not sin, wasn't it? Yes, but Jesus went beyond the symptoms to the cause. He would not heal the body and neglect the soul. He would not remedy a temporal condition, and leave an eternal condition untouched.

So He said, "your sins are forgiven." It was a wonderful announcement. Now—on this earth in this life—the man's sins were forgiven. He didn't have to wait till the day of Judgment. He had the present assurance of forgiveness. So do all who put their faith in the Lord Jesus.

The scribes quickly caught on to the significance of the statement. They were well enough trained in Bible doctrine to know that only God can forgive sins. Anyone who professed to forgive sins was therefore claiming to be God. Up to this point their logic was correct. But instead of acknowledging the Lord Jesus to be God, they accused Him in their hearts of blasphemy.

Jesus read their thoughts, a proof in itself of His supernatural power. He asked them this provocative question: "Which is easier, to say to the paralytic, 'Your sins are forgiven,' or to say, 'Rise, take up your bed and walk'?" Actually it is just as easy to *say* one as the other. But it is equally impossible, humanly speaking, to *do* the one as it is to *do* the other.

The Lord had already pronounced the man's sins forgiven. Yes, but had it really taken place? The scribes could not *see* the man's sin forgiven, therefore they would not believe. In order to demonstrate that the man's sins had really been forgiven, the Savior gave the scribes something they could see. He told the paralyzed man to get up, to carry his straw pad, and to walk. The man responded instantly. The people were amazed.

Only God can forgive sins.

They had never seen anything like it before. But did the scribes believe? No, they did not, in spite of the most overwhelming evidence. Belief involves the will, and they did not want to believe.

Levi Hears Christ's Call (2:13-14)

It was while He was teaching near the Sea of Galilee that Jesus saw Levi collecting taxes. We know Levi as Matthew, who later wrote the first gospel. He was a Jewish man, of course, but his occupation was very un-Jewish—he collected taxes for the despised Roman government. These men were not always noted for their honesty—in fact, they were looked down upon, like harlots, as the scum of society.

Yet it is to Levi's eternal credit that when he heard the call of Christ, he dropped everything to follow Him. May each of us be like him in instant and unquestioning obedience. It might seem like a great sacrifice at the time, but in eternity it will be seen as no sacrifice at all. "He is no fool who gives what he cannot keep, to gain what he cannot lose" (Jim Elliot).

The Friend of Sinners (2:15-17)

It seems that Levi arranged a banquet so he could introduce his friends to the Lord Jesus. Most of his friends were like himself—tax collectors and sinners. Jesus accepted the invitation to be present with them.

The scribes thought they had caught Him in a serious fault. Instead of going directly to Him, they went to His disciples and tried to undermine their confidence and loyalty. How was it that their Master ate and drank with tax collectors and sinners?

Jesus heard it and reminded them that healthy people don't need a doctor—only those who are sick.

The scribes thought they were well, therefore they did not recognize their need of the Great Physician. The tax collectors and sinners admitted their guilt and their need of help. Jesus came to call sinners, like them, not self-righteous people.

Is there a lesson in this for us? Yes, there is! We should not shut ourselves up in Christianized communities, having no contact with the world. Rather we should seek to befriend the ungodly with a view to introducing them to our Lord and Savior. In becoming a friend to sinners, we should not do anything that would compromise our testimony. Neither should we allow the unsaved to drag us down to their level. We should take the initiative in guiding the friendship into positive channels of spiritual helpfulness. It would be easier to isolate oneself from the wicked world, but Jesus didn't do it, and neither should His followers.

"Easier to decline an invitation to the house of the great than to go there and behave as the Son of God. Easier to refuse the things of sense than to use them without abuse. Easier to maintain a life of prayer away from the haunts of men, than to enter them maintaining constant fellowship with God in the unruffled depths of the soul. Nothing but the grace of the Holy Spirit can suffice for this" (F. B. Meyer).

The scribes thought they would ruin the Lord's reputation by calling Him a friend of sinners. But their intended insult has become an endearing tribute. All the redeemed gladly acknowledge Him as the friend of sinners, and will love Him eternally for it.

Who Should Fast? (2:18-22)

The disciples of John the Baptist and the Pharisees practiced fasting as a religious exercise. In the Old Testament it was instituted as an expression of deep sorrow. But it had lost much of its meaning and had become a routine ritual. They noticed that the disciples of Jesus didn't fast, and perhaps there was a tinge of envy and self-pity in their hearts when they asked the Lord for an explanation.

In reply He compared His disciples to companions of a bridegroom. He Himself was the Bridegroom, of course. As long as He was with them, there was no occasion for an outward demonstration of sorrow. But the day was coming when He would be taken away; then they would have occasion to fast.

Immediately the Lord added two illustrations to announce that a new era had arrived which was incompatible with the one that had passed.

The first illustration had to do with a new patch on old cloth. The new patch is made of cloth that has not been shrunk. If it is used to repair an old garment, it will inevitably shrink and something will have to give. Because the garment is made of older cloth, it will be weaker than the patch and will tear again wherever the patch is sewed to it.

Jesus was saying, in effect, that the Old Dispensation was like the old garment. God never intended Christianity to patch up Judaism; it was a new departure. The sorrow and sadness of the Old Era, which expressed itself in fasting, must give way to the joy of the New.

The second illustration involved new wine in old wineskins. The wineskins, made of leather, lost their power to stretch. If new wine was put into them, the pressure built up by the fermentation would burst the skins. The new wine typifies the joy and power of the Christian faith. The old skins depict the forms and rituals of Judaism. New wine needs new skins.

It was no use for John's disciples and the Pharisees to put the Lord's disciples under the bondage of sorrowful fasting, as it had been practiced. The joy and effervescence of the new life must be allowed to express themselves.

Christianity has always suffered from man's attempt to mix it with legalism. The Lord Jesus taught that the two are incompatible. Law and grace are opposing principles.

Lord of the Sabbath (2:23-28)

This incident illustrates what Jesus had just taught—the conflict between the traditions of Judaism and the liberty of the Gospel.

As the disciples went through the grain fields on the Sabbath, they picked some grain to eat. This didn't violate any law of God, but it did violate the traditions of the elders. According to their fantastic hair-splitting, the disciples had been guilty of breaking the Sabbath by reaping and perhaps even by threshing (rubbing the grain in their hands to remove the husks).

The Lord answered them by taking them back to an incident in the Old Testament. David had been anointed as king but he had been rejected, and instead of reigning, he was being hunted like a deer. One day when his provisions were gone, he entered into the house of God and used the the bread of the Presence to feed his men and himself. Ordinarily the the bread of the Presence was forbidden to any but the priests. Yet David was not rebuked by God for doing this.

Law and grace are opposing principles.

Why not? Because things were not right in Israel. As long as David was not given his rightful place as king, God allowed him to do what ordinarily would be illegal.

This was the case with the Lord Jesus. He had been anointed but He was not reigning. The very fact that His disciples had to pick grain as they traveled showed that things were not right in Israel. The Pharisees themselves should have been extending hospitality to Jesus and His disciples instead of criticizing them.

If David had actually broken the law by eating the the bread of the Presence, yet was not rebuked by God, how much more blameless were the disciples who, under similar circumstances, had broken nothing but the traditions of the elders.

It says in verse 26 that David ate the the bread of the Presence when Abiathar was high priest. According to 1 Samuel 21:1, Ahimelech was high priest at the time. However, Abiathar was his son, and it may be that his loyalty to David influenced the father to permit this unusual departure from the law.

Our Lord closed His discourse by reminding the Pharisees that the Sabbath was instituted by God for man's benefit, not for his bondage. He added that the Son of Man was Lord even of the Sabbath. It was He who had given the Sabbath in the first place. Therefore He had authority to decide what was permissible and what was forbidden on that day. Certainly the Sabbath was never intended to prohibit works of necessity or deeds of mercy.

Christians are not obligated to keep the Sabbath. That day was given to the nation of Israel. The distinctive day of Christianity is the Lord's Day, the first day of the week. However, it is not a day encrusted with legalistic do's and don'ts. Rather it is a day of privilege when, free from secular employments, believers may worship, serve and attend to the culture of their souls. For us it is not a question, "Is it wrong to do this on the Lord's

Day?" but rather "How may I best use this day to the glory of God, to the blessing of my neighbor, and to my own spiritual good?"

The Case of the Withered Hand (3:1-6)

Another test case arose on the Sabbath. As Jesus entered into the synagogue, He met a man with a withered hand. This raised the question, "Would Jesus heal him on the Sabbath?" If He did, the Pharisees would have a case against Him—or so they thought.

Imagine their hypocrisy and insincerity. They couldn't do anything to help this man, and they resented anyone who could. They sought some ground on which to condemn the Lord of life. If He healed on the Sabbath, they would rush in to the kill like a pack of wolves.

The Lord Jesus asked the man to come to Him. The atmosphere was charged with expectancy. Then He asked the Pharisees, "Is it lawful on the Sabbath to do good or to do harm, to save life or to kill?"

His question revealed the wickedness of the Pharisees' position. They thought it was wrong for Him to perform a miracle of healing on the Sabbath Day. But they didn't think it was wrong for them to plan His destruction on the Sabbath. He wanted to return a man's hand to usefulness. They wanted to kill Him.

No wonder they didn't answer His question! After an embarrassed silence, the Savior ordered the man to stretch out his hand. As he did so, full strength returned, the flesh filled out to normal size, and the wrinkles disappeared.

That was more than the Pharisees could take. They went out, contacted the Herodians, their traditional enemies, and plotted with them to destroy Jesus. It was still the Sabbath.

It was Herod who had brought about the death of John the Baptist. Perhaps his party would be equally successful in killing Jesus. This was the hope of the Pharisees.

Sermon by the Sea (3:7-12)

Jesus withdrew from the synagogue and went to the Sea of Galilee. The sea in the Bible often symbolizes the Gentile people. Therefore His action may have depicted His turning from the Jewish people to the Gentiles. A crowd gathered, not only from Galilee but from distant parts of the land. In fact, the crowd was so great that Jesus asked for a small boat so that He could push off from shore. Otherwise He would have been thronged by those who came for healing.

There were unclean spirits in the crowd too. Whenever they saw Him, they cried out that He was the Son of God. He ordered them to stop saying this. As already mentioned, He would not receive the witness of evil spirits. He did not deny that He was the Son of God, but He chose to control the time and manner for His being revealed as such.

Although Jesus had the power to heal, He did not heal everyone in the country. His miracles were performed only on those who came for help. That is the way it is with salvation. His power to save is sufficient for all, but efficient only for those who trust Him.

We learn from the ministry of the Savior that need does not constitute a call. There was need everywhere. Jesus depended on instructions from God the Father as to where and when to serve. So must we.

Twelve Disciples (3:13-19)

Faced with the task of world evangelization, Jesus called twelve disciples. There was nothing wonderful about the men themselves; it was their connection with Jesus that made them what they became.

They were young men when they heard His call and answered. In his book, *The Life and Teaching of Jesus Christ*, James E. Stewart has this splendid commentary on the youth of the disciples:

"Christianity began as a young people's movement. In thinking of Jesus and His disciples, that is the first fact to make quite clear. Unfortunately, it is a fact which Christian art and Christian preaching have too often obscured. But it is quite certain that the original disciple band was a young men's group. Most of the apostles were probably still in their twenties when they went out after Jesus.

> **Christ's power to save is sufficient for all, but efficient only for those who trust Him.**

"Jesus Himself, we should never forget, went out to His earthly ministry with the 'dew of (His) youth' upon Him (Psalm 110:3—this psalm was applied to Jesus first by Himself and then by the apostolic church). It was a true instinct that led the Christians of a later day, when they drew the likeness of their Master on the walls of the catacombs, to portray Him, not old and weary and broken with pain, but as a young shepherd out on the hills of the morning. The original version of Isaac Watt's great hymn was true to fact:

When I survey the wondrous cross
Where the young prince of Glory died.

"And no one has ever understood the heart of youth in its gaiety and gallantry and generosity and hope, its sudden loneliness and haunting dreams and hidden conflicts and strong temptations, no one has understood it nearly so well as Jesus. And no one ever realized more clearly than Jesus did that the adolescent years of life, when strange dormant thoughts are stirring and the whole world begins to unfold, are God's best chance with the soul. When Jesus and youth come together, deep calls to deep. There is an immediate, instinctive feeling of kinship, and everything that is fine and noble and pure in youth bows down in admiration and adoration before Him.

"It is not surprising then, that Christianity entered the world as a young people's movement. When we study the story of the first twelve, it is a young men's adventure we are studying. We see them following their Leader out into the unknown, not knowing very clearly who He is or why they are doing it or where He is likely to lead them; but just magnetized by Him, fascinated and gripped and held by something irresistible in the soul of Him, laughed at by friends, plotted against by foes, with doubts sometimes growing clamorous in their own hearts, until they almost wished they were well out of the whole business; but still clinging to Him, coming through the ruin of their hopes to a better loyalty and earning triumphantly at last the great name the Te Deum gives them, 'The glorious company of the apostles.' It is worth watching them, for we too may catch the infection of their spirit and fall into step with Jesus."

There was a threefold purpose behind the call of the Twelve:

1. That they might be with Him.
2. That He might send them forth to preach.
3. That they might have authority over demons.

First, there was to be a time of training. There was to be preparation in private before preaching in public. Here is a basic principle of service. We must spend time with God before we move out as His representatives.

Secondly, they were sent forth to preach. Proclamation of the Word of God was their basic method of evangelism. That must always be central. Nothing must be allowed to push it into a subordinate place.

Finally, they were given supernatural power. The casting out of demons would attest to people that God was speaking through the apostles. At that time, the Bible had not yet been completed. Miracles were the credentials of God's messengers. Today people have access to the complete Word of God; they are responsible to believe it without the proof of miracles.

One name stands out in the name of the apostles, that of Judas Iscariot. There is mystery connected with his being chosen as an apostle, then turning out to be the betrayer of our Lord. One of the greatest heartaches in Christian service is to see one who was bright, earnest and apparently devoted, later turning his back on the Savior and going back to the world that crucified Him.

Eleven proved true to the Lord, and through them He turned the world upside down. They reproduced themselves in ever-widening circles of outreach, and in one sense, we today are the continuing fruit of their service. There is no way of telling how far-reaching our influence for Christ may be.

The Unforgivable Sin (3:20-30)

Jesus returned from a mountain, where He had called His disciples, to a Galilean home. The crowd gathered in such numbers that He and His apostles were kept too busy to eat.

When His family heard of His activities, they felt that He was out of His mind. They sought to take Him away. Doubtless they were ashamed of Him. They had a religious Fanatic in the family and they were embarrassed by His zeal. Even our Lord's relations did not understand Him. His life was so unworldly that it could not be measured by the ordinary standards. Here they could account for His unconquerable zeal only by concluding that He was insane. We hear much of the same kind of talk in modern days when some devoted follower of Christ utterly forgets self in love for his Master. People say, 'He must be insane!' They think every person is crazy whose religion kindles into any sort of unusual fervor, or who grows more earnest than the average Christian in work for the Master; some of Paul's friends thought he was crazy when he went sweeping over land and sea to carry the Gospel to every city. But his answer was that he wasn't crazy, 'for the love of Christ controls us.'

"That is a good sort of insanity. It is a sad pity that it is so rare. If there were more of it there would not be so many unsaved souls dying under the very shadow of our churches; it would not be so hard to get missionaries and money to send the Gospel to the dark continents; there would not be so many empty pews in our churches; so many long pauses in our prayer-meetings; so few to teach in our Sunday schools. It would be a glorious thing if all Christians were beside themselves as the Master was, or as Paul was.

"It is a far worse insanity which in this world never gives a thought to any other world; which, moving continually among lost men, never pities them, nor thinks of their lost condition, nor puts forth any effort to save them. It is easier to keep a cool head and a colder heart, and to give ourselves no concern about perishing souls; but we are our brothers' keepers, and no malfeasance in duty can be worse than that which pays no heed to their eternal salvation" (J. R. Miller, *Daily Reading* for June 6).

It is always true that a person who is on fire for God seems to be deranged to their contemporaries. The more like Christ we are, the more we too will be exposed to the sorrow of being misunderstood by relatives and friends. If we set out to make a fortune, people will cheer us. If we are fanatics for Jesus Christ, they will jeer us.

The scribes did not think He was insane. They accused Him of casting out demons by the power of Beelzebub, prince of the demons. (The name Beelzebub in Syriac means lord of filth.) This was a most serious and blasphemous charge. First Jesus refuted it, then pronounced the doom of those who made it.

If He were casting out demons by Beelzebub, then Satan would be working against himself. The devil would be frustrating his own purposes. His aim is to control people through demons, not to free them from demons. Beelzebub would never fight against himself. A

> **The more like Christ we are, the more we too will be exposed to the sorrow of being misunderstood.**

kingdom, a house, or a person divided against themselves cannot endure. Continued survival depends upon internal cooperation, not antagonism.

The scribes' accusation was therefore preposterous. In fact, the Lord Jesus was doing the very opposite of what they said. His miracles signified the downfall of Satan rather than his prowess. That is what the Savior meant when He said, "No one can enter a strong man's house and plunder his goods, unless he first binds the strong man. Then indeed he may plunder his house" (v. 27).

Satan is the strong man. The house is his dominion; he is the god of this age. His goods are the people over whom he holds sway. Jesus is the One who binds Satan and spoils his house.

At Christ's second advent, Satan will be bound and cast into the bottomless pit for one thousand years. When the Savior cast out demons during His ministry on earth, it was a forecast of His eventual complete binding of the devil.

In verses 28-30, the Lord pronounced the doom of the scribes who were guilty of the unforgivable sin. The scribes had accused Jesus of casting out demons by demonic power. Actually it was by the power of the Holy

Spirit that He did it. They therefore said that the Holy Spirit was a demon. This is blasphemy against the Holy Spirit.

All kinds of sin can be forgiven, but this particular sin has no forgiveness. It is an eternal sin.

Can people commit this particular sin today? Probably not. It was a sin, which people committed when Jesus was on earth performing miracles. Since He is not on earth today in the same sense, casting out demons, the same possibility of blaspheming the Holy Spirit does not exist.

People who worry that they have committed the unpardonable sin have not done so. The very fact that they are concerned indicates that they are not guilty of blasphemy against the Holy Spirit.

Natural and Spiritual Ties (3:31-35)

Mary, the mother of Jesus, came with His brothers to talk with Him. The crowd prevented their getting to Him, so they sent word that they were waiting outside for Him. When the messenger told Him that His mother and His brothers wanted Him, He looked out over the crowd and announced that His mother and His brothers were those who do the will of God.

Several lessons emerge from this for us:

- First of all, the words of the Lord Jesus show the proper attitude that we should take toward Mary. He did not dishonor her as His natural mother, but He did say that spiritual relationships take precedence over natural ones. The passage clearly teaches that it was more to Mary's credit to do the will of God than to be His mother.
- Secondly, the passage disproves the idea that Mary remained a virgin the rest of her life. Jesus had brothers. He was Mary's firstborn, but she gave birth to other sons and daughters afterwards (see Matthew 13:55; Mark 6:3; John 2:12; 7:3, 5, 10; Acts 1:14; 1 Corinthians 9:5; Galatians 1:19).
- Jesus put God's interests above natural ties. To His followers, He still says today: "Whoever does the will of God, he is my brother and sister and mother."
- The passage reminds us that believers are bound by stronger cords to fellow-Christians than they are to blood-relations when those relatives are unsaved.
- Finally, it emphasizes the importance He places on doing the will of God. Do I meet the standard? Am I His mother or brother?

LESSON 2 EXAM

Use the exam sheet at the back of the course to complete your exam.

1. **Jesus assured the paralyzed man that his sins were forgiven as**
   A. an evidence of His own sympathy and cooperation.
   B. an unconditional blessing for him to enjoy then and there.
   C. a promise for the future only.
   D. a substitute for the healing of his body.

2. **The scribes' reaction to Jesus' claim to forgive sins was**
   A. the proclamation of His power to the people.
   B. belief by some of them.
   C. to accuse Him of blasphemy in their hearts.
   D. first amazement then belief.

3. **The disciples of Jesus did not fast because**
   A. as long as Jesus was with them there was no occasion to fast.
   B. fasting tends to degenerate into a mere routine ritual.
   C. fasting was an Old Testament custom abolished by the Lord.
   D. Jesus Himself never fasted.

4. **The Lord's illustration of the new patch and the old cloth teaches the truth that**
   A. Christianity is a new form of Judaism.
   B. it is possible for Judaism and Christianity to coexist.
   C. a new era has arrived, incompatible with the one that has passed.
   D. the Gospel can patch up the consequences of sin.

5. **The idea that law and grace are opposing principles and that Christianity has nothing in common with legalism is**
   A. foreign to the New Testament.
   B. inherent in the teachings of Jesus.
   C. a Pauline idea superimposed on the teachings of Jesus.
   D. incompatible with the teachings of Jesus.

6. **From the healing of the man with the withered hand we learn that**
   A. despite outward religious observance it is possible for a person to be hypocritical.
   B. the Pharisees did actually have a case against Christ.
   C. the Herodians were more inclined towards Christ than the Pharisees.
   D. politics and religion should be combined for the cause of Christ.

7. **The calling of the Twelve impresses us with the fact that**
   A. the primary appeal of the Gospel is to the wealthy and the educated.
   B. ignorant and uneducated men have difficulty understanding Christ's call.
   C. Christianity entered the world as a young people's movement.
   D. youth and Christianity have little or nothing in common.

8. **Miracles are not necessary today as credentials for the Lord's servants because**
   A. they never were necessary as credentials.
   B. men are different now and no longer believe in miracles.
   C. God expects men to believe His written Word without miracles.
   D. miracles are today in the realm of science.

9. **The unforgivable sin**
   A. can probably be committed today only by those ignorant of the claims of Christ.
   B. is the sin of attributing to demonic power that which the Lord Jesus did in the power of the Holy Spirit.
   C. has probably been committed by all those who imagine they have done so.
   D. is the sin of murder.

10. **The Lord's words to Mary**
    A. show that He dishonored her.
    B. show the proper attitude that we should have towards Mary.
    C. show that it was more important for her to be His mother than to do God's will.
    D. teach us that natural relationships are as important as spiritual ones.

What Do You Say?

What do we learn from the fact that Christ chose the disciples from a variety of backgrounds?

LESSON 3

Parables of the Kingdom

Mark 4

Four Types of Soil (4:1-20)

Again Jesus went to teach by the seaside. Again the crowd made it necessary for Him to use a boat as His pulpit, just a short way from the beach. And again He taught spiritual lessons from the world of nature about Him. He could see spiritual truth in the natural realm. It is there for all of us to see.

The first parable has to do with:

1. The Sower
2. The Seed
3. The Soil

- *The Path Soil.* The ground was too hard for the seed to get down into it. Birds came and ate the seed.
- *The Rocky Ground.* A thin layer of dirt covered a bed of rock. Shallowness of earth prevented the seed from taking deep root.
- *The Thorny Ground.* Thorn bushes cut the seed off from nourishment and sunlight, thus choking it.
- *The Good Ground.* The seed fell into deep, fertile soil where conditions were favorable. Some seeds produced thirty kernels, some sixty and some one hundred.

The Parable Explained

When the disciples were with Him alone, they asked Him why He spoke in parables. He said that to them was given the secret or the mystery of the kingdom of God.

What did He mean by the mystery of the kingdom of God? A mystery in the New Testament is a truth hitherto unknown, and one that can only be known through special revelation. The mystery of the kingdom of God is that:

- The Lord Jesus was rejected when He offered Himself as King to Israel.
- A period of time would intervene before the kingdom would be literally set up on earth.
- During the interim, it would exist in spiritual form. All who acknowledge Christ as King would be in the kingdom, even though the King Himself was absent.
- The Word of God would be sown during the interim period with varying degrees of success. Some people would actually be converted, but others would be only nominal believers. All would be in the kingdom in its outward form, but only the genuine ones would enter the kingdom in its inner reality.

Why was this truth presented in parables? The explanation is given in verses 11 and 12. God reveals His family secrets to those whose hearts are open, receptive and obedient. On the other hand, the truth is deliberately hidden from those who reject the light that is given to them. These are the people Jesus referred to as "those outside."

The words of verse 12 may seem harsh and unfair to the casual reader: "They may indeed see but not perceive, and may indeed hear but not understand, lest they should turn and be forgiven." But we must remember the tremendous privilege that these people had enjoyed. The Son of God had taught in their midst and had performed many mighty miracles before them. Instead of acknowledging Him as the true Messiah, they were even now rejecting Him. Because they had spurned the Light of the world, they would be denied the light of His teachings. Henceforth they would see His miracles, yet not understand the spiritual significance. They would hear His words; yet not appreciate the deep lessons in them.

There is such a thing as hearing the Gospel for the last time. It is possible to sin away the day of grace. People do drift beyond redemption point. There are men and women who have refused the Savior and who will never again have the opportunity to repent and be forgiven. They may hear the Gospel but it falls on hardened ears and an insensible heart. We say, "Where there's life, there's hope," but the Bible speaks of some who are alive and yet who are beyond hope (Hebrews 6:4-6, for example).

Going back to the parable of the sower, the Lord Jesus asked the disciples how they could expect to understand more involved parables if they could not understand this simple one (v. 13).

The Savior did not identify the sower. It could be Himself or those who preach as His representatives. The seed, He said, is the Word. The various types of soil represent human hearts and their receptivity to the Word, as follows:

- *The path soil.* This heart is hard. The person is stubborn and unbroken. They say a determined "No" to the Savior's invitation. Satan, pictured by the birds, snatches away the Word. The person is unmoved and untroubled by the message. They are indifferent and insensible to it thereafter.
- *The rocky ground.* This pictures a person who makes a superficial response to the Word. Perhaps in the emotion of a fervent Gospel appeal, they make a profession of faith in Christ. But it is just a mental assent. There is no real commitment of the person to Christ. They receive the Word with joy; it would be better if they received it with deep repentance and contrition. They seem to go on brightly for a while, but when tribulation or persecution arise because of their profession, they decide that the cost is too great and they abandon the whole thing. They claim to be a Christian as long as it is popular to do so, but persecution exposes their unreality.
- *The thorny ground.* These people also make a promising start. To all outward appearances, they seem to be true believers. But then they become preoccupied with business, with worldly worries, with the lust to become rich. They lose interest in spiritual things, until finally they abandon any claim to be Christians at all.
- *The good ground.* Here there is a definite acceptance of the Word, cost what it may. These people are truly born again. They are loyal subjects of Christ, the King. Neither the world, the flesh nor the devil can shake their confidence in Him.

Even among the good ground hearers, there are varying degrees of fruitfulness. Some bear thirtyfold, some sixtyfold and some one hundredfold. What determines the degree of productivity? The life that is most productive is the one that obeys the Word promptly, unquestioningly, and joyfully.

The Responsibility of Those Who Hear (4:21-25)

The lamp here (candle, KJV) represents the truths that the Lord imparted to His disciples. These truths were not to be put under a bushel or a bed, but put out in the open for all to see. The bushel doubtless represents business, which if allowed, will steal time that should be given to the things of the Lord. The bed may speak of comfort, ease or even laziness, all of which are enemies of evangelism.

Jesus spoke to the multitude in parables. The underlying truth was hidden. It was made secret at that time. But the divine intention was that the disciples should take those hidden truths and explain them to willing hearts wherever they found them (v. 22).

Verse 22 might also mean, however, that the disciples should serve in constant remembrance of a coming day of manifestation. At that time, it will be seen if business or self-indulgence were allowed to take precedence over testimony for the Savior.

The seriousness of these words is indicated by Jesus' admonition: "If anyone has ears to hear, let him hear."

And then the Savior added another serious warning to His disciples: "Pay attention to what you hear." Every time we learn some truth from the Word of God, we are responsible to obey it and then to share it with others. To receive divine truth puts us under obligation to do something about it.

> **To receive divine truth puts us under obligation to do something about it.**

If I hear some command from the Word of God but fail to obey it, I will not be able to pass it on to others. What gives power and scope to teaching is when people see the truth in the preacher's life.

Whatever we measure out to others in the way of sharing the truth comes back to us with compound interest. The teacher usually learns more in the preparation of a lesson than the pupils. Then too the future reward will be greater than our poor, puny expenditure.

Every time we acquire fresh truth and allow it to become real in our lives, we are sure to be given more truth. On the other hand, failure to respond to truth results in a loss of that which was previously acquired.

Growth and Harvest (4:26-29)

His next parable is found only in the Gospel of Mark. It can be interpreted in at least two ways:

1. It may picture the Lord Jesus casting seed on the earth during His public ministry, then returning to heaven. The seed begins to grow—mysteriously, imperceptively but invincibly. From a small beginning, there develops a harvest of true believers. When the proper time arrives, the grain will be harvested and taken to the heavenly garner.
2. It may be intended as an encouragement to disciples of the Lord Jesus. Their responsibility is to sow the seed. They may sleep by night and rise by day (v. 27), knowing that God's Word will not return to Him void, but will accomplish what He has intended it to do. By a mysterious and miraculous process, quite apart from man's strength and skill, the Word takes effect in human hearts, producing fruit for God. People plant and water but God gives the increase.

The difficulty with this last interpretation lies in verse 29. Only God can put forth the sickle when harvest time has arrived. But in the parable the same man who sows the seed puts in the sickle when the grain is ripe.

The Mustard Seed (4:30-34)

This parable pictures the growth of the kingdom from a beginning as small as a mustard seed to a tree or bush big enough for the birds to roost in. The kingdom started off with a small, persecuted minority. Then it became more popular and was embraced by governments as the state religion. While this growth was spectacular, it was not healthy. Much of the growth represented people who paid lip service to the King but who had never been truly converted.

"As long as the church wore scars," said Vance Havner, "they made headway. When they began to wear medals, the cause languished. It was a greater day for the church when Christians were fed to the lions than when they bought season tickets and sat in the grandstand."

The mustard bush therefore could picture professed Christendom, which has become a roosting place for all kinds of false teachers. It is the outward form of the kingdom as it exists today.

Verses 33 and 34 introduce us to a very important principle in teaching. Jesus taught the people "as they were able to hear it." He built upon the knowledge they had, permitting time for them to assimilate one lesson before giving them the next. He was conscious of the capacity of His hearers, and did not glut them with more instruction than they could absorb (see also John 16:12; 1 Corinthians 3:2; Hebrews 5:12).

Although His general teaching was in parables, He expounded them to His disciples in private. Those who sincerely desire light are given it.

Peace, Be Still (4:35-41)

At evening of the same day, Jesus and His disciples started across the Sea of Galilee toward the eastern shore. They had not made any advance preparations. Other boats followed. Then suddenly a violent windstorm arose. Huge waves threatened to swamp the boat.

Jesus was sleeping in the stern of the boat. The frantic disciples awoke Him, rebuking Him for His seeming lack of concern for their safety. The Lord arose and rebuked the wind and the waves.

> **Those who sincerely desire light are given it.**

The calm was immediate and complete. Then Jesus briefly chided His followers for fearing and not trusting. They were stunned by the miracle. Even though they knew who Jesus was, they were impressed afresh by the power of One who could control the elements.

The incident reveals the humanity and the deity of the Lord Jesus. He slept in the stern of the boat; that's His humanity. He spoke and the sea was calm; that's His deity. It demonstrates His power over nature, as previous miracles showed His power over diseases and demons.

Finally, it encourages us to go to Jesus in all the storms of life, knowing that the boat can never sink when He is in it.

> Thou art the Lord who slept upon the pillow,
> Thou art the Lord who soothed the furious sea,
> What matter beating wind and tossing billow
> If only we are in the boat with Thee?
> Hold us in quiet through the age-long minute
> While Thou art silent, and the wind is shrill;
> Can the boat sink while Thou, Dear Lord, art in it?
> Can the heart faint that waiteth on Thy will?
>
> —Amy Carmichael

LESSON 3 EXAM

Use the exam sheet at the back of the course to complete your exam.

1. **The Parable of the Sower emphasizes the fact that**
   A. most people's hearts are hard like rocks and impenetrable.
   B. shallowness is the most prevalent condition of the human heart.
   C. the Gospel message will ultimately be universally accepted.
   D. only some of those who receive the message produce fruit for God.

2. **A New Testament "mystery" is**
   A. something hidden and unknowable this side of eternity.
   B. a truth until this time unknown and knowable only through special revelation
   C. a secret revealed only to a special class of Christian initiates.
   D. something revealed in the Old Testament but only understood in the New Testament.

3. **Which of the following is not a part of "the mystery of the kingdom"?**
   A. The Lord Jesus was rejected by Israel as King.
   B. The kingdom of God is forever exclusively a spiritual kingdom in the hearts of men.
   C. All who acknowledge Christ as King will be in the kingdom.
   D. Nominal believers are in the kingdom in its outward form.

4. **The Lord Jesus presented truth in parables**
   A. in order to help "them that are without" grasp the truth He was teaching.
   B. in order to make it understandable to a select class of true believers.
   C. in order to deliberately hide the truth from those who reject the light that is given to them.
   D. in order to show that we can fully find God apart from the Scriptures.

5. **The significant lesson we learn from the Lord's words "that seeing they may see, and not perceive …" is that**
   A. only a few people will ever get an opportunity to hear the Gospel.
   B. "where there's life there's hope".
   C. it is possible to drift beyond the point of redemption in continually hardening one's heart.
   D. only intelligent and educated people can believe and understand the Gospel.

6. **The path soil represents those who, when confronted with the Savior's invitation**
   A. accept it but give up their profession of faith when the way becomes hard and difficult.
   B. accept it but become preoccupied with either wealth or worry and abandon their profession.
   C. would like to accept it but are afraid of being persecuted.
   D. remain unmoved and untroubled by the message and become indifferent and insensible to it.

7. **The lamp, in the parable of the bushel basket, represents**
   A. the truths imparted by the Lord to His disciples.
   B. the individual testimony of believers quenched by fear.
   C. the corporate testimony of the Church often endangered by compromise with the world.
   D. the nation of Israel, to be extinguished as God's vehicle for imparting truth to the nations.

8. **The parable of the growth of the seed and the harvest (Mark 4:26-29) can mean**
   A. that the seed sown by the Lord during His earthly ministry begins small during His absence in heaven, and will ultimately be gathered in by the Lord.
   B. that the Lord's disciples are responsible to sow the seed and, although they may sleep by night and rise by day, the mysterious, miraculous process of growth will go on in men's hearts, as God's Word takes effect, apart from human strength and skill.
   C. either of the above ways although A is to be preferred because the sowing and the reaping is done by the same one.
   D. either A or B although B is best because it is in keeping with the Lord's parable of the sower, the seed and the soil.

9. **The parable of the mustard seed could picture**
   A. the Church triumphant on earth, growing and spreading until all nations shelter in its various branches.
   B. professed Christendom (outward form of Kingdom today) which has become a haven for all kinds of false teachers.
   C. the smallness and insignificance of the Church that nevertheless provides shelter for all who fly to it for refuge.
   D. the development and growth of divine truth in a receptive human heart.

10. **The miracle of Christ's stilling the storm**
   A. is recorded in Mark's Gospel only and even there is open to much question as to its authenticity.
   B. must be understood allegorically, as indicating the Lord's power to still the storms of life and give us peace, rather than literally.
   C. is actually a parable rather than a miracle and is intended to picture Christ's power over all the forces of earth and heaven.
   D. demonstrates both the humanity and deity of the Lord Jesus and shows His power over nature.

What Do You Say?

The Lord stilled the storm on the lake. What does this incident teach you about the Lord Jesus?

LESSON 4

Miracles of Healing

Mark 5

My Name Is Legion (5:1-20)

The country of the Gerasenes was on the east side of the Sea of Galilee. There Jesus met a demon-possessed man who was unusually violent. He was a terror to society. Every effort to restrain him had failed. He lived among the tombs and on the mountains, yelling continually and gashing himself with sharp stones.

When the demoniac saw Jesus, he first acted respectfully, then complained bitterly. How like the sinner he was!

> **The sinner is a slave—helpless, desolate, and self-tortured.**

"The sinner is a slave—helpless, desolate, self-tortured, often yearning for deliverance, yet unwilling to be set free. How true and terrible a picture is this—a man bowed in adoration, petition and faith, and yet hating, defiant and fearing; a double personality, longing for liberty and yet clinging to passion" (Scripture Union Notes).

The exact order of events is not clear, but may have been as follows:

- The demoniac performed an act of reverence to the Lord Jesus (v. 6).
- Jesus ordered the unclean spirit to come out of him (v. 8).
- The spirit, speaking through the man, acknowledged who Jesus was, challenged His right to interfere and ordered Jesus with an oath to stop tormenting him (v. 7).
- Jesus asked the man's name. It was Legion, signifying he was indwelt by many demons (v. 9). This apparently does not contradict verse 2 where it says he had an unclean spirit (singular).
- Perhaps it was the spokesman for the demons who begged permission to enter into a herd of pigs (vv. 10-12).

- Permission was granted with the result that two thousand pigs raced down the mountainside and drowned in the sea (v. 13).

The Lord has often been criticized for causing the destruction of these pigs. Several points should be noted:

- He did not cause this destruction; He permitted it. It was Satan's destructive power that destroyed the pigs.
- There is no record of the owners finding fault. Perhaps they were Jewish people for whom the raising of pigs was forbidden.
- The soul of the man was worth more than all the pigs in the world.
- If we knew as much as Jesus knew, we would have acted exactly the same way He did.

The people who witnessed the destruction of the swine ran back to the city with the news. A large crowd returned to find the ex-demoniac sitting at Jesus' feet clothed and sane. The people were afraid. Someone has said, "They were afraid when He stilled the tempest on the sea, and now in a human soul." Those who had been witnesses rehearsed the whole story to the newcomers. It was too much for the populace: they urged Jesus to leave the area. This and not the destruction of the pigs is the shocking part of the incident. Christ was too costly a guest!

Evangelism begins at home.

"Countless multitudes still wish Christ far from them for fear His fellowship may occasion some social or financial or personal loss. Seeking to save their possessions, they lose their souls" (Selected).

As Jesus was about to leave by boat, the healed man asked to accompany Him. It was a worthy request and evidenced his new life, but Jesus sent him to his home as a living witness of God's great power and mercy. The man obeyed, carrying the good news to Decapolis, an area embracing ten cities.

This is a standing order for all who have experienced the saving grace of God: "Go home to your friends and tell them how much the Lord has done for you, and how He has had mercy on you." Evangelism begins at home!

Urgent Request of Jairus (5:21-24)

Back upon the western shore of blue Galilee, the Lord Jesus was soon in the center of a crowd. A frenzied father came running up to Him. It was Jairus, one of the rulers of the synagogue. His little daughter was dying. Would Jesus please go and lay His hands on her so that she might be healed?

The Lord, of course, responded and started for the home. A crowd followed Him, thronging Him.

It is interesting that immediately following the statement of the crowd's thronging Him, we have an account of faith touching Him.

Healed by Faith (5:25-34)

A distracted woman intercepted Jesus on the way to the home of Jairus. It was a seeming interruption, but our Lord was neither annoyed nor ruffled. This speaks to our hearts. How do *we* react to interruptions?

"I think I find most help in trying to look on all interruptions and hindrances to work that one has planned out for oneself as discipline, trials sent by God to help one against getting selfish over one's work. Then one can feel that perhaps one's true work—one's work for God—consists in doing some trifling haphazard thing that has been thrown into one's day. It is not waste of time, as one is tempted to think, it is the most important part of the work of the day—the part one can best offer to God. After such a hindrance, do not rush after the planned work; trust that the time to finish it will be given sometime, and keep a quiet heart about it" (Choice Gleanings Calendar).

This woman had suffered with chronic bleeding for twelve years. The many doctors she went to had apparently used some drastic forms of treatment, had drained her financial resources, and had left her worse rather than better. When hope of recovery was all but gone, someone told her about Jesus. She lost no time finding Him, then eased her way through the crowd till she touched the lower border of His garment. Immediately the bleeding stopped and she felt completely well.

Her plan then was to slip away quietly, but the Lord would not let her miss the blessing of publicly acknowledging Him as her Savior. He had been aware of an outflow of divine power when she touched Him; it cost Him something to heal her. So He asked who had touched Him. He knew the answer, of course, but asked in order to bring her forward in the crowd.

The disciples thought the question was silly. So many people were jostling Him continually. Why ask who touched Him? But there is a difference between the touch of physical nearness, and the touch of desperate faith. It is possible to be ever so near Him without trusting Him, but it is impossible to touch Him by faith without His knowing it and without being healed.

Fearful and trembling, the woman came forward, fell down before Him and made her first public confession of Jesus. Then He spoke words of assurance to her soul.

Open confession of Christ is of tremendous importance. Without it there can be little growth in the Christian life. Also it is as we take our stand boldly for Him that He floods our souls with full assurance of faith. The words of the Lord Jesus (v. 34) confirmed the fact that she had been healed physically, but they also no doubt included the great blessing of soul salvation as well.

Raising of Jairus' Daughter (5:35-43)

By this time, some messengers had arrived with the sad news that Jairus' daughter had died. There was no need to bring the Teacher, therefore. The Lord graciously reassured Jairus, then took Peter, James, and John to the house. They were met by the unrestrained weeping that is characteristic of eastern homes in times of sorrow; some of it is done by hired mourners, of course.

Souls that have known the throb of new life need to be fed.

When Jesus assured them that their weeping and wailing were needless because the child was not dead but sleeping, their tears turned to scorn. Undaunted, He took the immediate family to the motionless child and taking the girl by the hand, He said in Aramaic, "Little girl, I say to you, arise." Immediately the twelve-year-old girl got up on her feet and walked. The relatives were stunned, and doubtless delirious with joy.

The Lord forbade their publicizing the miracle. He was not interested in the popular acclaim of the masses. He must resolutely press on to the cross.

If the girl had actually died, then this chapter illustrates the power of Jesus over demons, disease, and death. Not all Bible scholars agree that she was dead. Jesus said she was not dead but sleeping. It may be that she was in a deep coma. He could just as easily have raised her from the dead, but He would not take credit for doing so if she were only unconscious.

We should not overlook the closing words of the chapter, "He... told them to give her something to eat." In spiritual ministry, this would be known as "follow-up work." Souls that have known the throb of new life need to be fed. One way a disciple can manifest his love for the Savior is by feeding His sheep.

LESSON 4 EXAM

Use the exam sheet at the back of the course to complete your exam.

1. **The violent, demon-possessed man of the Gerasenes**
   A. was abusive to the Lord Jesus up to the moment the Lord commanded the evil spirits to come out of him.
   B. asked the Lord Jesus His name.
   C. illustrates the personality of the sinner who at once loves and hates his bondage.
   D. was regarded as a holy man by the neighboring villagers because he was indwelt by spirits.

2. **The villagers, who had been terrorized by the demon-possessed man of the Gerasenes,**
   A. were grateful to the Lord Jesus for healing the demoniac.
   B. requested the Lord to stay and teach them more of the things of God.
   C. remind us of those who are afraid that personal involvement with the Lord Jesus will cost them.
   D. brought their sick to Jesus that He might heal them too.

3. **This healed Gerasene demoniac**
   A. became a disciple of the Lord Jesus and followed Him from that day forward.
   B. was challenged by the Lord to "come, take up thy cross and follow me" but, instead, went home to his family.
   C. went back to the tombs and to his old way of life eventually.
   D. wanted to follow Jesus but was sent back by Him to be a witness at home.

4. **When Jairus approached the Lord Jesus about his dying daughter:**
   A. Jesus ignored him.
   B. The crowd told Jairus to go away.
   C. Jesus immediately started to go with Jairus.
   D. Jesus deliberately waited where he was.

5. **As the crowd thronged around Jesus, a women with a flow of blood touched Him. We learn**
   A. that Jesus will always respond when someone reaches out in faith.
   B. that Jesus was annoyed at the interruption.
   C. that Jairus was very upset at the delay.
   D. that the woman was healed gradually.

6. **The women planned to**
   A. announce her healing to the world.
   B. slip away quietly.
   C. go to see the Priest.
   D. go and tell the doctors about her healing.

7. **From the story of the healing of the woman with the issue of blood we can learn**
   A. that no case is too hard for the Lord Jesus.
   B. that it is possible to be very near to the Lord and yet not to reach out by faith and touch Him.
   C. those whose lives have been miraculously touched by the Savior ought to publicly confess Him.
   D. all the above lessons.

8. **When the Lord Jesus arrived at the home of Jairus:**
   A. The daughter had recovered.
   B. The daughter appeared to be dead.
   C. The crowd thought the girl was well.
   D. Jairus was not there.

9. **The Lord did not want the people to publicize the healing. This was because**
   A. it was not a great miracle.
   B. Jesus was not interested in popular acclaim, but pressing on to the cross.
   C. it would not help Him get to the next town.
   D. Jesus was tired.

10. **In this lesson we have seen the Lord Jesus dealing triumphantly with**
    A. the natural elements, disease, and demon possession.
    B. a man, a woman, and a child.
    C. a Jew, a Samaritan, and a Greek.
    D. a sinner, a backslider, and a devoted disciple.

What Do You Say?

What impression has this study of the Lord Jesus made upon you personally so far?

LESSON 5

Only Believe

Mark 6

Without Honor in His Own Country (6:1-6)

Jesus returned to Nazareth with His disciples. This was "His hometown" where He had worked as a carpenter. On the Sabbath He taught in the synagogue. The people were amazed. They could not deny the wisdom of His teaching or the wonder of His miracles. But there was a deep unwillingness to acknowledge Him as the true Son of God. They thought of Him as a carpenter, the son of Mary, whose brothers and sisters were still there. Had He returned to Nazareth as a mighty conquering Hero, they might have accepted Him more readily. But He came in lowly grace and humility. This stumbled them.

It was then that Jesus observed that a prophet is generally given a better reception away from home. His relatives and friends were too close to him to appreciate his person or ministry. No place is harder to serve the Lord than at home.

The Nazarenes themselves were a despised people. A popular proverb was: "Can any good thing come out of Nazareth?" Yet these social outcasts looked down upon the Lord Jesus when He appeared in their midst. What a commentary on the pride and unbelief of the human heart!

It is clear from verse 3 that Mary did not remain a virgin after Jesus was born. She later gave birth to sons and daughters.

Unbelief largely hindered the work of the Savior in Nazareth. He healed a few sick folks, but that was all. The unbelief of the people amazed Him. "Such unbelief as this has immense consequences for evil. It closes the channels of grace and mercy, so that only a trickle gets through to human lives in need" (J. G. Miller).

Again Jesus tasted the loneliness of being misunderstood and slighted. Many of His followers have shared this sorrow. Often the servants of the Lord appear in a very humble guise. Are we able to look beyond outward appearances and recognize true spiritual worth?

The Lord was not daunted by His second rejection in Nazareth. He went about the surrounding villages, teaching the Word.

> His foes might hate, despise, revile;
> His friends unfaithful prove.
> Unwearied in forgiveness still
> His heart could only love.

Disciples Sent Forth (6:7-13)

The time had come for the disciples to launch out. They had been under the matchless tutelage of the Savior; now they would go forth as heralds of a glorious message.

He sent them out two by two. The preaching would thus be confirmed in the mouths of two witnesses. Also there would be strength and mutual help in traveling together. Finally, the presence of two might be helpful in cultures where moral conditions were low.

Next He gave them authority or power over unclean spirits. This is worth noting. It is one thing to cast out demons; it is another thing to confer on others the power to cast out demons. Only God can do this.

If our Lord's kingdom were of this world, He would never have given the instructions that follow in verses 8-11. They are the very opposite of what the average worldly leader would give.

- The disciples were to go forth without provisions. No bread, no wallet, no money. They were to trust Him for the supply of their needs. Theirs was to be a life of faith. They were allowed to take sandals and a staff, the latter perhaps for protection against animals. One coat would be enough. Certainly no one would envy the disciples for their possessions. People would not be attracted to Christianity by the prospect of becoming wealthy. And whatever power the disciples would have must come from God; they were totally cast upon Him. "They

Theirs was to be a life of faith.

were to be sent forth in meanest circumstances, yet representatives of the Son of God, invested with His power."

It was the way the Master went!
Shall not His servant tread it still?

- They were to accept hospitality wherever it was offered to them, and they were to remain there until they left the area. This instruction prevented their shopping around for more comfortable lodgings. After all, their mission was to preach the message of One who pleased not Himself, who was not self-seeking. They were not to compromise the message by seeking luxury, comfort or ease.
- If a place rejected the disciples and refused the message, they were not obligated to remain. To do so would be casting pearls before swine. In leaving, the disciples were to shake the dust off their feet, symbolizing God's rejection of those who reject His beloved Son.

Although some of the instructions were of a temporary nature and were later withdrawn by the Lord Jesus (Luke 22:35-36), yet they embody lasting principles for the servant of Christ in every age.

The disciples went forth, preached repentance, cast out demons, anointed the sick with oil, and healed them. The anointing with oil, we believe, was a symbolic gesture, picturing the soothing, alleviating power of the Holy Spirit.

Herod Mistakes Jesus for John (6:14-16)

When news reached Herod that a miracle-worker was traveling through the land, he immediately concluded that it was John the Baptist, risen from the dead. Others said it was Elijah or one of the other prophets, but Herod was quite convinced that the man whom he had beheaded had risen.

The terrible pangs of conscience were stabbing Herod for what he had done. John the Baptist had been a voice from God. Herod had silenced that voice. Now his conscience was catching up with him. He would learn that the way of the transgressor is hard.

How John Had Been Killed (6:17-29)

The narrative now switches back to the time when John had been executed. The Baptizer had reproved Herod for entering into an unlawful marriage with his brother's wife. His wife Herodias became furious and vowed to take revenge. But Herod respected John as a holy man and thwarted her efforts.

Finally her chance came. At Herod's birthday party, with local celebrities attending, Herodias arranged for her daughter to dance. This so pleased Herod that he promised to give the girl anything up to the half of his kingdom. Prompted by her mother, she asked for the head of John the Baptist on a platter.

Herod was trapped. Against his own desires and better judgment, he granted the request. Sin had woven its web around him, and the vassal king was victimized by an evil woman and by a sensual dance.

When John's faithful disciples heard what had happened, they claimed his body and buried it, then went and told Jesus.

Feeding the Five Thousand (6:30-44)

This miracle is found in all four Gospels. It took place at the beginning of the third year of His public ministry. The disciples had just returned to Capernaum from their first preaching mission (see verses 7-13). Perhaps they were flushed with success, perhaps weary and footsore. At any rate, the Lord Jesus recognized their need for rest and quiet, so He took them by boat to a secluded area on the shore of the Sea of Galilee.

> **Are people an intrusion to us, or the objects of our love?**

We often hear the words, "Come away by yourselves to a desolate place and rest a while" used to justify luxurious vacations for world-bordering Christians. William Kelly once wrote: "It would be well for us if we needed thus to rest more; that is to say, if our labors were so abundant, our self-denying efforts for the blessing of others were so continual, that we could be sure that this was the Lord's word for us."

A crowd followed the Lord and His disciples by taking the land route along the shore of the lake. Jesus pitied the people. They were wandering around without a spiritual guide, hungry and defenseless. So He began to teach them.

As the day wore on, the disciples became restless about the crowd—so many people and nothing to eat. They urged the Lord to send them home. The same crowd that drew out the compassion of the Savior proved an annoyance to the disciples. Are people an intrusion to us, or the objects of our love?

Jesus turned to the disciples and said, "You give them something to eat." The whole thing seemed preposterous—five thousand men, plus women and children, and nothing but five loaves and two fishes—and God.

In the miracle that followed, the disciples would see a picture of how the Savior would give Himself to be the bread of life for a starving world. His body would be broken that others might have eternal life. In fact, the words used are highly suggestive of the Lord's Supper that commemorates His death: He took; He blessed; He broke; He gave.

The disciples would also learn precious lessons in connection with their service for Him.

- The Lord has infinite power to provide the needs of His people. We need never doubt His ability to "spread a table in the wilderness."
- The perishing world seems too great to feed. How can the world ever be evangelized? Jesus says, "You give them something to eat." If we give Him what we have, He can multiply it in blessing to multitudes. He uses what we have, however trivial it may seem.
- He handled the work in a systematic, orderly way by seating the crowd in groups of hundreds and fifties.
- He blessed and broke the loaves and fishes. Unblest by Him, they would never have availed. Unbroken they would have been utterly insufficient. "The reason we are not more freely given to men is that we are not yet properly broken" (Selected).
- Jesus did not distribute the food Himself. He allowed His disciples to do this. His plan is to feed the world through His people.
- There was sufficient for all. If believers today would put everything above current necessities into the work of the Lord, the whole world could hear the Gospel in this generation.
- The fragments that were left over (twelve basketsful) were more than He started with. God is a bountiful Giver. Yet notice that nothing was wasted. The surplus was gathered up.
- Disciples of the Lord Jesus should never doubt His power to supply their needs. If He can feed five thousand men with five loaves and two fishes, He can provide for His trusting servants under any circumstances. They can labor for Him without worry as to where their food is coming from. If they seek first the kingdom of God and His righteousness, every need will be supplied.

Jesus Walks on the Water (6:45-52)

Not only can the Savior provide for the sustenance of His servants; He can provide for their safety as well.

After sending the disciples back to the west shore of the lake by boat, Jesus Himself went up into a mountain to pray. In the darkness of the night, He saw them rowing hard against a strong wind. He went to their assistance, walking on the water. At first, they were terrified, thinking it was a ghost. Then He spoke reassuringly to them and boarded the boat. The wind ceased immediately.

The account closes with the comment: "They were utterly astounded, for they did not understand about the loaves, but their hearts were hardened." This seems almost unconnected with what has gone before. But the thought seems to be this: Even after seeing the power and greatness of the Lord in the miracle of the loaves, they still did not realize that nothing was impossible for Him. They shouldn't have been surprised to see Him walking on the water. It was no greater a miracle than the one they had just witnessed. Lack of faith produced hardness of heart and dullness of spiritual perception.

The church has always seen in this miracle a picture of the present age and its close:

- Jesus on the mountain represents Christ in His present ministry in heaven, interceding for His people.
- The disciples represent His servants being buffeted by the storms and trials of life.
- Soon the Savior will return to His own, deliver them from danger and distress and guide them safe to the heavenly shore.

Continued Ministry of Healing (6:53-56)

Back on the west side of the lake, the Lord was besieged with sick folks. No matter where He went, people carried needy cases to Him on mats. The market places became improvised hospitals. All they wanted was to get close enough to Him to touch the hem of His garment. Whenever they did this, healing power flowed to them.

LESSON 5 EXAM

Use the exam sheet at the back of the course to complete your exam.

1. **The people of His home town regarded Jesus as**
    A. a conquering hero.
    B. coming from a despised town.
    C. the true Son of God.
    D. the carpenter.

2. **The Lord was unable to do many great things in Nazareth because**
    A. the people refused to believe in Him.
    B. He was there so seldom.
    C. John the Baptist came from Nazareth and the people idolized him.
    D. the people were blinded by idolatry.

3. **The Lord sent the disciples out in pairs**
    A. that their testimony might be confirmed by two witnesses.
    B. because there is strength and mutual help in traveling together.
    C. as protection against moral temptation.
    D. for all the above reasons.

4. **The Lord's instructions to His disciples, as they went forth with the message,**
    A. was designed to attract people to Christianity by the affluence and material success of the disciples.
    B. cast them on Him to supply their material needs.
    C. were very similar to those any great leader would give to his followers when launching a sales campaign.
    D. included instructions to take an adequate supply of provisions since they could not expect strangers to support them.

5. **The instructions given by Jesus to His disciples as they went forth with the message**
   A. were permanent and apply, in every detail in this age.
   B. were so idealistic as to be unworkable in practice and have all been withdrawn for Christian workers.
   C. were temporary but, although they were later withdrawn, still embody lasting principles for the Lord's servants.
   D. were designed, really, for those who will herald His message during the days of the coming Great Tribulation.

6. **Herod was convinced that Jesus was John the Baptist, risen from the dead because**
   A. his conscience was gnawing him for having silenced the convicting voice of John.
   B. he was by nature a very superstitious person.
   C. Jesus was performing the same kind of miracles John had performed.
   D. John's disciples were broadcasting this very story.

7. **Herod had murdered John the Baptist because**
   A. he hated the man and feared his influence.
   B. he disagreed violently with John's teachings.
   C. he was carried away in a moment of passion with a desire to please two evil women and to save face before his friends.
   D. he considered it a politically opportune thing to do.

8. **The spiritual lesson of the miracle of the feeding of the five thousand is that**
   A. most people are easily satisfied.
   B. the Lord Jesus is the bread of life for a starving world.
   C. mass evangelism is far more important than personal soul winning.
   D. the social gospel is the true gospel.

9. **What would be a legitimate lesson to draw from a study of the miraculous feeding of the five thousand?**
    A. The people followed the lad's example, and they all gave until all were filled.
    B. The work of the Gospel should be entrusted only to those who are able to use sound management techniques.
    C. Since we have such a limited source of supply, economy should be the first consideration in the Lord's work.
    D. God's work, done through human channels, is effective when the Lord blesses it.

10. **The Lord's walking on the water**
    A. was a far greater miracle than His feeding of the five thousand.
    B. pictures the present age and its close.
    C. must be understood only in an allegorical sense.
    D. can be fully explained by modern science.

What Do You Say?

What does the Lord's feeding of the 5,000 tell us about His ability to meet our needs?

Belief and Unbelief

Mark 7, 8

Tradition vs. the Law of God (7:1-23)

The scribes and Pharisees were Jewish religious leaders who had built up a vast system of traditions. These traditions were enforced rigidly. They were so interwoven with the law of God that they had acquired almost equal authority with the Scriptures. But in some cases they actually contradicted the Scriptures or weakened the law of God.

The religious leaders delighted in imposing the rules and the people accepted them meekly, satisfied with a system of rituals without reality.

Here in Mark chapter seven, we find the Pharisees and scribes criticizing Jesus because His disciples ate with unwashed hands. This doesn't mean that the disciples didn't wash their hands before they ate. Rather it means that they didn't go through the elaborate ritual prescribed by tradition. Unless, for instance, they washed up to the elbows, they were considered ceremonially defiled. If they had been in the market place, they were supposed to take a ceremonial bath. This complex system of washing extended even to the dipping of pots and pans.

"There are some who build their whole lives around negatives … . They [the Pharisees] came all the way from Jerusalem to meet Him, and their life attitudes were so negative and faultfinding that all they saw was unwashed hands. They couldn't see the greatest movement of redemption that had ever touched our planet—a movement that was cleansing the minds and souls and bodies of men. All they saw was a ritualistic infringement. Their big eyes were opened wide to the little and marginal, and blind to the big. So history forgets them, the negative—forgets them except as a background for this impact of the positive Christ. They left a criticism;

He left a conversion. They picked flaws; He picked followers." (E. Stanley Jones, *Growing Spiritually*, Nashville: Abingdon Press, 1953, p. 109.)

Jesus quickly pointed out the hypocrisy of such behavior. The people were just what Isaiah had predicted. They outwardly professed great devotion to the Lord, but inwardly they were corrupt. By elaborate external rituals, they pretended to worship God, but had substituted their own traditions for the doctrines of the Bible.

Instead of recognizing the Word of God as the sole authority in all matters of faith and morals, they evaded or explained away the clear demands of the Scripture by their traditions.

The disciples had failed to observe the *traditions of men*. Their accusers had forsaken the *commandments of God*. Jesus made this very clear.

Then He singled out an example of how tradition had made void the law of God. One of the Ten Commandments given through Moses demanded that children should honor their parents. The death penalty was decreed for anyone who spoke evil of his father or mother (which apparently meant failing to care for them in their need).

> **The Word of God is the sole authority in all matters of faith.**

But a Jewish tradition had arisen known as Corban, which meant "given" or "dedicated." Here is how it worked! Suppose that certain Jewish parents were in great need. Their son had money to care for them, but he didn't want to do it. All he had to do was say "Corban," implying that his money was dedicated to the Lord or to the temple. This relieved him of any further responsibility to support his parents. He might keep the money indefinitely and use it in business. Whether it ever was turned over to the temple was not important.

"Just consider what an issue this was. A man sees his father and mother in want; he has received in earthly goods that which would relieve them, but the tradition-mongers have invented a plan to benefit religion so-called at the cost of filial duty. If one said, 'Corban,' the duty was totally changed; and that which would have been due to the parent must now be devoted to the priest. No matter what the need of father and mother, that word 'Corban' stopped all action of heart or conscience. The leaders had devised the scheme to secure property for religious purposes and to quiet persons from all trouble of conscience about the Word of God.

"But the Judge and Lord of all meets this at once. Who had given them authority to say, 'It is Corban'? Where had God warranted such a practice? And who were they that dared to substitute their thoughts for the Word of God? It was God who called on man to honor his parents, and who denounced all slight done to them. Yet here were men violating, under cloak of religion, both these commandments of God! This tradition of saying 'Corban,' the Lord treats not only as a wrong done to the parents, but as a rebellious act against the express commandment of God." (William Kelly, *An Exposition of the Gospel of Mark*, London: C.A. Hammond, 1934, p. 105.)

Beginning at verse 14, the Lord made the revolutionary pronouncement that it is not what goes into a person's mouth that defiles them (such as food eaten with unwashed hands) but what comes out of person (such as traditions that set aside God's Word).

Even the disciples were mystified by this. Brought up under the teachings of the Old Testament, they had always considered that certain foods like pork, rabbit, and shrimp were unclean and would defile them. Jesus now plainly stated that a person was not defiled by what went into them. In a sense, this signaled the end of the legal dispensation.

The last part of verse 19 in the Revised Version reads: "This He said, making all meats clean." In other words, the distinction between clean and unclean foods that prevailed in the Old Testament was now abolished.

Someone has said, "This He said, making all foods clean … and making all hypocrisy unclean."

It's what comes out of man's heart that defiles him, things like sexual immorality, theft, murder, deceit, railing and pride. And in the context, the thought is that human tradition should be listed here too. The tradition of Corban was tantamount to murder. Parents could die of starvation before this wicked vow could be broken.

One of the great lessons we learn from this passage is that we must constantly test all teaching and all tradition by the Word of God, obeying what is of God and rejecting what is of mankind. At first a person may teach and preach a clear, scriptural message. This gives them acceptance among Bible—believing people. Having gained this popular acceptance, they begin to add some human teaching. By this time, their devoted followers have come to feel that he/she can do no wrong. So they follow them blindly, even if their message blunts the sharp edge of the Word or waters down its clear meaning.

> It's what comes out of person's heart that defiles them.

It was in this way that the scribes and Pharisees had gained authority as teachers of the Word. In reality, they were now nullifying the intent of the Word. The Lord Jesus had to warn the people that it is the Word that accredits people, not people who accredit the Word. The great touchstone must always be, "What says the Word?"

Daughter of a Gentile Woman Healed (7:24-30)

In the preceding incident Jesus pronounced all foods clean. Here He demonstrates that Gentiles are no longer common or unclean.

Jesus now traveled northwest to the coast of Tyre and Sidon, also known as Syrophoenicia. He tried to enter a home incognito, but His fame had preceded Him and His presence was soon known. A Gentile woman came to Him, asking for help for her demon-possessed daughter.

We emphasize the fact that she was a Gentile, not a Jewish woman. The Jewish people, of course, were God's chosen earthly people. They occupied a place of distinct privilege with God. He had made wonderful covenants with them. He had committed the Scriptures to them. He had dwelt with them in the tabernacle, then in the temple.

By contrast, the Gentiles were aliens from the commonwealth of Israel, strangers from the covenants of promise, without Christ, without hope, without God in the world (Ephesians 2:11-12).

The Lord Jesus came primarily to the nation of Israel. He presented Himself as King to that nation. The Gospel was first preached to the house of Israel. It is important to see this in order to understand His dealings with the Syrophoenician woman.

When she asked Him to cast the demon out of her daughter, He seemed to rebuff her. He said that the children (Israelites) should first be filled, that it was not proper to take the children's food and throw it to dogs (Gentiles). His answer was not a refusal. He said, "Let *the children* be fed first."

This might sound like a harsh reply. Actually it was a test of her repentance and faith. He was simply saying that His ministry at that time was directed primarily to the Jewish people. Not being an Israelite, she had no claim upon Him or upon His benefits. This was true. But would she acknowledge it?

> We must constantly test all teaching and all tradition by the Word of God.

She did. She said, in effect, "True, Lord. I am only a Gentile dog. But I notice that puppies have a way of eating crumbs that children drop from the table. That's all I ask for—some crumbs left over from your ministry to the Jewish people."

This faith was remarkable. The Lord rewarded it instantly by healing the daughter at a distance. When the woman got home, the girl was fully recovered.

Deaf Man Healed (7:31-37)

From the coast of the Mediterranean, our Lord returned to the east coast of the Sea of Galilee—the area known as Decapolis. There an incident took place that is recorded only in Mark's Gospel.

Interested friends brought to Him a deaf man who also had an impediment in his speech. Maybe this impediment was caused by a physical deformity or perhaps by the fact that he never heard sounds correctly and so could not reproduce them correctly. At any rate, he illustrates the fact that the sinner is deaf to the voice of God and therefore unable to speak to others about Him.

> **Christ's sigh expressed His grief over the suffering which sin has brought on mankind.**

Jesus first took the man aside privately. Then He put His fingers in his ears, spat and touched his tongue. Thus by a sort of sign language, He told the man that He was about to open his ears and unloose his tongue. Next Jesus looked up to heaven, indicating that the power He used was from God. His sigh expressed His grief over the suffering which sin has brought on mankind. Finally He said "Ephphatha," the Aramaic word for "Be opened."

The man obtained normal hearing and speech immediately. The Lord asked the people not to publicize the miracle, but they disregarded His instructions. Disobedience can never be justified, no matter how well-meaning the persons might be.

The spectators were amazed by His wonderful works. They said, "He has done all things well. He even makes the deaf hear and the mute speak." They did not know the full truth of what they said. Had they lived on this side of Calvary, they would have said it with deeper conviction and feeling.

> 'Ere since our souls have learned His love,
> What mercies has He made us prove!
> Mercies which all our praise excel!
> Our Jesus hath done all things well!

Four Thousand Fed (8:1-9)

This miracle seems strangely similar to the feeding of the five thousand, yet notice the following differences:

The Five Thousand	The Four Thousand
1. The people were Jewish, see John 6:14-15.	1. The people were probably Gentiles (they lived in Decapolis).
2. The multitude had been with Jesus one day (Mark 6:35).	2. This crowd had been with Him three days (Mark 8:2).
3. Jesus used five loaves and two fishes (Matthew 14:17).	3. He used seven loaves and a few fishes (Mark 8:5, 7).
4. Five thousand men, plus women and children were fed (Matthew 14:21).	4. Four thousand men, plus women and children were fed (Matthew 15:38).
5. The surplus filled twelve hand baskets (Matthew 14:20).	5. The surplus filled seven wicker baskets or hampers (Mark 8:8).

The less Jesus had to work with, the more He accomplished and the more He had left over (at least as far as the number of baskets).

In the preceding chapter, we saw crumbs falling from the table to a Gentile woman. Here we see a multitude of Gentiles being fed abundantly. Erdman comments: "The first miracle in this period intimated that crumbs of bread might fall from the table for the needy Gentiles; here they may be an intimation that Jesus, rejected by His own people, is to give His life for the world, and is to be the living Bread for all nations."

There is a danger in treating incidents like the feeding of the four thousand as repetition without significance. We should approach Bible study with the conviction that every word of Scripture is filled with spiritual truth, even if we can't see it at our present state of understanding.

Pharisees Demand Sign from Heaven (8:10-13)

From the area of Decapolis, Jesus and the disciples crossed the Sea of Galilee to the west side, to a place called Dalmanutha (Magdala in Matthew 15:39).

There the Pharisees waited for Him, demanding a sign from heaven. Their blindness and boldness were enormous. Standing in front of them was the greatest Sign of all—the Lord Jesus Himself. He was truly a Sign who had come from heaven, but they had no appreciation for Him. They heard His matchless words, they saw His wonderful miracles, they came in contact with an absolutely sinless Man—God manifest in the flesh. Yet in their blindness they asked for a sign from heaven.

No wonder the Savior sighed deeply! If any generation in the history of the world had been privileged, it was the Jewish generation of which those Pharisees were a part. Yet they were blind to the clearest evidence that the Messiah had appeared, and they asked for a miracle in the heavens rather than on earth. Jesus said, "There won't be any more signs. You've had your chance." He entered the boat again and sailed eastward.

Beware of Leaven (8:14-21)

During the journey the disciples remembered that they had forgotten to take bread along. Jesus was still thinking of His encounter with the Pharisees, however, when He warned them against the leaven of the Pharisees and the leaven of Herod.

Leaven in the Bible is a consistent type of evil, spreading slowly and quietly and affecting everything it touches. The leaven of the Pharisees combines the thoughts of hypocrisy, ritualism, self-righteousness and bigotry. The Pharisees made great outward pretensions of sanctity but inwardly they were corrupt and unholy.

The leaven of Herod may include the ideas of skepticism, immorality and worldliness. The Herodians were conspicuous for these sins.

The disciples completely missed the point of Jesus' warnings. All they could think of was food. So He directed nine rapid questions to them. The first five reproved them for their obtuseness. The last four rebuked them for worrying about the supply of their needs as long as He was with them. Had He not fed five thousand with five loaves, leaving twelve handbasketsful over? Yes! Had He not fed four thousand with seven loaves, leaving seven hampersful over? Yes, He had. Then why did they not understand that He was abundantly able to supply the needs of a handful of disciples in a boat? Didn't they realize that the Creator and Sustainer of the universe was in the boat with them?

Blind Man Receives Sight (8:22-26)

This miracle is found only in the Gospel of Mark. There are several interesting questions that arise in connection with it.

- First, why did Jesus lead the man out of the village before healing him?
- Why didn't He heal by simply touching the man? Why use such an unconventional means as spittle?

- Why didn't the man receive perfect sight immediately? (This is the only cure in the Gospels that took place in stages.)
- Why did Jesus forbid the healed man to tell about the miracle in the village?

Our Lord is sovereign in His ways of acting. He can do as He pleases without giving an account of His actions to us. There was a valid reason for everything He did, even though we might not perceive it.

Every case of healing is different, just as is every case of conversion. Some gain remarkable spiritual sight as soon as they are converted. Others see dimly at first, then later enter into full assurance of salvation.

Peter's Confession of Faith (8:27-30)

The last two paragraphs of this chapter bring us to the high water mark of the training of the Twelve. It was necessary for the disciples to have a deep, personal appreciation of who Jesus is. As soon as they came to know this for themselves, He was able to share with them the pathway ahead and to invite them to follow Him in a life of devotion and sacrifice. This passage brings us to the heart of discipleship. It is perhaps the most neglected area in Christian thought and practice today.

Jesus and the disciples sought solitude in the far north. On the way to Caesarea Philippi, He opened the subject by asking what public opinion said of Him. In general, people were acknowledging Him to be a great man—on the same level with John the Baptist, Elijah or other prophets. But mankind's honor is in reality dishonor. If Jesus is not God, then He is a deceiver, a madman or a legend. There is no other possibility.

Then the Lord pointedly asked the disciples for their evaluation of Him. Peter promptly declared Him to be the Christ, that is, the Messiah or the Anointed One. Intellectually Peter had known this before. But something had happened in his life so that now there was a profound, personal, moving conviction. Life could never be the same again. Peter could never be satisfied with a self-centered existence. If Christ was the Messiah, then Peter must live for Him in total abandonment. He could say:

> I have seen the vision
> And for self I cannot live.
> Life is worse than worthless
> Unless all I give.

Jesus Foretells His Death and Resurrection (8:31-38)

Up to this point we have watched the Servant of Jehovah in a life of incessant service for others. We have seen Him hated by His enemies and misunderstood by His friends. We have seen a life of dynamic power, of moral perfection, of utter love and humility.

But the path of service to God leads on to suffering and death. So the Savior now told the disciples plainly that He must (1) suffer; (2) be rejected; (3) be killed; (4) rise again.

For Him the path to glory would lead first to the cross and the grave. "The heart of service would be revealed in sacrifice." (F. W. Grant).

Peter could not accept the idea that Jesus the Messiah would have to suffer and die; that was contrary to his image of the Messiah. Neither did he want to think that his Lord and Master would be slain by His foes. So he rebuked the Savior for even suggesting such a thing. It was then that Jesus said to Peter, "Get behind me, Satan! For you are not setting your mind on the things of God, but on the things of man." Not that Jesus was accusing Peter of being Satan, or even of being indwelt by Satan. But it was as if He said, "You are talking like Satan would. He always tries to discourage us from wholly obeying God. He tempts us to take an easy path to the Throne." The words of Peter were Satanic in their origin and contents. It was this that caused the Lord to be so indignant.

> **The path of service to God leads on to suffering and death.**

"What was it that so roused our Lord? The very snare to which we are all so exposed: the desire of saving self; the preference of an easy path to the cross. Is it not true that we naturally like to escape trial, shame, and rejection; that we shrink from the suffering which doing God's will, in such a world as this, must ever entail; that we prefer to have a quiet, respectable path in the earth—in short, the best of both worlds? How easily one may be ensnared into this! Peter could not understand why the Messiah should go through all this path of sorrow. Had we been there, we might have said or thought even worse. Peter's protest was not without strong human affection. He heartily loved the Savior too. But, unknown to himself, there was the unjudged spirit of the world." (Kelly, William, *An Exposition of the Gospel of Mark*, London: C.A. Hammond, 1934, p. 136.)

Note in verse 33 that Jesus first looked on His disciples, then rebuked Peter, as if to say, "If I do not go to the cross, how can these, my disciples, be saved?"

Then Jesus said, in effect, "I am going to suffer, bleed and die in order that people might be saved. If you would come after me, you must deny every selfish impulse, deliberately strike out on a pathway of reproach, suffering and death, and follow me. You may have to forsake personal comforts, social enjoyments, earthly ties, grand ambitions, material riches and even life itself."

When we read words like these, it makes us wonder! How can we really believe that it is all right for us to live in luxury and ease? How can we spend our lives accumulating wealth? How can we justify the materialism, the selfishness, the coldness of our hearts? His words call us to lives of self-denial, surrender, suffering, and sacrifice.

There is always the temptation to save our lives—to live comfortably, to provide for the future, to make one's own choices, with self as the center of everything. There is no surer way of losing one's life.

Christ calls us to pour out our lives for His sake and the Gospel's. We should dedicate ourselves to Him, spirit, soul and body. He asks us to spend and be spent in His holy service, laying down our lives, if necessary, for the evangelization of the world. That is what is meant by losing our lives. There is no surer way of saving them.

> **Christ calls us to live lives of self-denial, surrender, suffering, and sacrifice.**

Even supposing that a believer could gain all the wealth of the world during his lifetime, what good would it do him? He would have missed the opportunity of using his life for the glory of God and the salvation of the lost. It would be a bad bargain. Our lives are worth more than all the world has to offer. Shall we use them for Christ or for self?

Even as He spoke to His young disciples, our Lord realized that some of them might be stumbled in the path of discipleship by the fear of shame. So He reminded them that those who seek to avoid reproach because of Him will suffer a greater shame when He returns to earth in power and great glory. What a thought! Soon our Lord is coming back to the earth. He will not come in humiliation then, but in His own personal glory and in the glory of His Father, with the holy angels. It will be a scene of dazzling splendor. He will then be ashamed of those who are ashamed of Him now.

Notice the words "whoever is ashamed of me and of my words in this adulterous and sinful generation." May this speak to our hearts. How incongruous to be ashamed of the sinless Savior in a world that is characterized by unfaithfulness and sinfulness.

Should we to gain the world's applause
Or to escape its harmless frown,
Refuse to countenance Thy cause,
And make Thy people's lot our own,
What shame would fill us in that day,
When Thou Thy glory wilt display.

LESSON 6 EXAM

Use the exam sheet at the back of the course to complete your exam.

1. **The systems of tradition practiced by the scribes and Pharisees included**
   A. substituting a system of rituals to take the place of reality.
   B. substituting human tradition for the doctrines of the Bible.
   C. professing devotion to the Lord while being inwardly corrupt.
   D. All of these practices were and are Pharisaical.

2. **The Jewish tradition known as "Corban"**
   A. enabled a man to avoid his family responsibilities under the cover of religion.
   B. enabled a man to avoid paying his tithes and taxes.
   C. made it possible for a man to sin with impunity under cover of a priestly indulgence that could be purchased.
   D. was defended by the Lord as a legitimate way of discharging one's religious obligations.

3. **At first the Lord did not help the Syropheonician woman because**
   A. she was a Gentile and therefore outside the scope of His mercy and care.
   B. He wanted to test her repentance and faith.
   C. she failed to address Him with proper respect and give Him His appropriate title.
   D. it would have clashed with the racial and religious prejudices of the Jews.

4. **After Jesus healed the man who was deaf and who had an impediment in his speech, the people said of Him**
   A. "This man receives sinners and eats with them."
   B. "Never man spoke like this man."
   C. "He has done all things well."
   D. "Behold, how He loved him."

5. **Which of the following is not a point of contrast between the feeding of the five thousand and the feeding of the four thousand?**
   A. In one case the people fed were Jews, in the other they were probably Gentiles.
   B. In the one case Jesus gave the broken bread to the disciples to distribute, in the other case He distributed it.
   C. In one case He used five loaves, in the other case, seven loaves.
   D. In one case the multitude had been with Jesus one day, in the other, three days.

6. **When the Lord spoke of the leaven of the Pharisees**
   A. He was using leaven as a type of the Gospel which would work for the good of the world.
   B. they understood at once the significance of His statement.
   C. they understood Him to be warning them against skepticism, and immorality.
   D. He was actually warning them against hypocrisy, ritualism, and self-righteousness.

7. **The blind man, healed by Jesus, said "I see men as trees, walking." This means**
   A. the man's faith was too weak for him to be completely healed.
   B. men, in many ways, resemble trees.
   C. every case of healing is different, just as every case of conversion is.
   D. Mark's reporting was second hand and therefore subject to error since he was not an eye witness.

8. **To evaluate the Lord Jesus as being equal with other religious teachers is to**
   A. honor Him greatly.
   B. dishonor Him completely.
   C. properly evaluate His Person and His work.
   D. accept Peter's evaluation of Him.

9. **Peter's reaction, when the Lord spoke of His plan of going to the cross,**
   A. was completely devoid of human sympathy.
   B. showed how greatly Peter had grown in spiritual perception.
   C. prompted the Lord to tell Peter that he was indwelt by Satan.
   D. revealed a sincere desire on his part to save Christ from suffering.

10. **The Lord's teaching to believers regarding losing their lives means that**
    A. believers should value their lives and not endanger them.
    B. only those disciples who are wholly dedicated die young.
    C. a life wholly abandoned to Christ, to the extent of laying that life down, is a life truly saved.
    D. those who do not give up everything for Christ will be eternally lost.

What Do You Say?

What is your response to Jesus' teaching about self-denial and following Him?

LESSON 7

Two Worlds

Mark 9

The Transfiguration (9:1-13)

The Lord Jesus had just laid before the disciples the pathway of reproach, suffering and death that He was to take. He had invited them to follow Him in lives of sacrifice and self-renunciation.

Now He gives them the other side of the picture. Though discipleship would cost them dearly in this life, it would be rewarded with glory by and by.

The Lord began by saying that some of the disciples would not taste of death till they saw the kingdom of God coming with power. He was referring to Peter, James, and John. On the Mount of Transfiguration they saw the kingdom of God in power. The transfiguration was a foregleam of Christ's glorious reign on earth.

Christ invited them to follow Him.

The argument of the passage is that anything we may suffer for Christ's sake now will be abundantly repaid when He returns in glory, and His servants appear with Him in glory.

The conditions which prevailed on the Mount foreshadow the millennial reign of Christ.

1. Jesus was transfigured. This means that dazzling splendor radiated from His Person. Even His clothes were shining–whiter than any bleach could make them.

    During His first advent, the glory of Christ was veiled. He came in humiliation, a Man of Sorrows, and acquainted with grief. But when He comes again, He will come in glory. No one will mistake Him then. He will be visibly the King of kings and Lord of lords.

2. Moses and Elijah were there. They represent:
   - Old Testament saints, or
   - The law (Moses) and the prophets (Elijah), or
   - Saints who have died, and those who have been translated
3. Peter, James, and John were there. They may represent New Testament saints in a general way, or those who will be alive when the kingdom is set up.
4. Jesus was the central Person. Peter's ill-considered suggestion of making three tabernacles was rebuked by the cloud and the voice from heaven. In all things Christ must have the preeminence. He will be the glory of Emmanuel's land.

The cloud may have been the Shekinah or glory cloud that stayed in the Holy of Holies in the tabernacle and temple in Old Testament times. It was the visible expression of God's presence. The voice was the voice of God the Father, acknowledging Christ as His beloved Son.

When the cloud was lifted, the disciples saw no man but only Jesus. It was a picture of the unique, glorious, and preeminent place He will have when the kingdom comes in power, and which He should have in the hearts of His followers at the present time.

As they came down from the mountain, He told them not to discuss what they had seen until after He had risen from the dead. This latter point puzzled them greatly. Perhaps they still did not quite take it in that He was to be slain and rise again. But more probably they wondered about the expression "risen from the dead." As Jewish people they knew the truth of resurrection, that is, the simple fact that all would be raised. But Jesus was speaking of a selective resurrection. He would be raised from among the dead ones. In other words not all would be raised when He arose. This is a truth that is found only in the New Testament.

Faith in the living God is always rewarded. Nothing is too difficult for Him.

The disciples had another problem. They had just had a preview of the kingdom. But hadn't Malachi predicted that Elijah would come as a fore-runner of the Messiah, beginning the restitution of all things, and paving the way for setting up His universal reign (Malachi 4:5)? Where was Elijah? Would he come, as the scribes said he would?

Jesus answered in effect, "Yes, it is true that Elijah must come first. But a more important and immediate question is this: 'Don't the Old Testament

Scriptures predict that the Son of Man is to endure great sufferings and be treated with contempt?' As far as Elijah is concerned, Elijah did come (in the person and ministry of John the Baptist) but people treated him exactly as they wanted to—just as people treated Elijah. The death of John the Baptist was an advance token of what they would do to the Son of Man. They rejected the forerunner; they will reject the King."

Demon-Possessed Boy Healed (9:14-29)

The disciples were not permitted to remain on the mountain-top of glory. In the valley below was groaning, sobbing mankind. A world of need lay at their feet.

When Jesus and the three disciples reached the base of the mountain, an animated discussion was going on between the scribes, the crowd, and the other disciples. As soon as the Lord appeared, the conversation broke up and the crowd rushed to Him. "What are you arguing about with them?" He inquired.

Then the story came out. A distraught father excitedly told the Lord about his son, possessed with a dumb spirit. The demon dashed the child to the ground, made him grind his teeth, and foam at the mouth. These violent convulsions were causing the child to waste away. The father had asked the disciples to do something, but they were powerless.

Jesus plaintively chided the disciples for their unbelief. Had He not already given them power to cast out demons? How long would He have to be with them before they would utilize the authority He had given them? How long would He have to put up with lives of powerlessness and defeat?

As they brought the child to the Lord the demon induced a particularly serious fit. The Lord tenderly asked the father how long this had been going on. It was from childhood, he explained. These spasms had often thrown the child into the fire and into the lake. There had been narrow escapes from death.

Then the father asked the Lord to please do something if He could. It was a heart-rending cry, wrung from years of desperation.

The Savior told him that it was not a question of His ability to heal, but rather of the father's ability to believe. Faith in the living God is always rewarded. Nothing is too difficult for Him.

The father expressed the paradox of faith and unbelief that God's people in all ages have experienced. "I believe; help my unbelief!" We want to believe; yet we find ourselves filled with doubt. We hate ourselves for this inward contradiction. It is so unreasonable. Yet we seem to fight it in vain.

When Jesus ordered the demon to leave the child, there was another terrible spasm, then the little body relaxed as if dead. The Savior raised him up and restored him to his father.

Later when our Lord was alone with the disciples in the house, they asked Him why they hadn't been able to do it. He replied that certain miracles require prolonged periods of prayer.

Which of us are not faced at times in our Christian service with a sense of utter defeat and frustration? We have labored tirelessly and conscientiously, yet there has been no evidence of the Spirit of God working in power. We too hear the Savior's words reminding us, "Nothing can drive out this kind of thing except prayer" (Phillips).

Jesus Again Predicts His Passion (9:30-32)

Our Lord's visit to the area of Caesarea Philippi had come to an end. Now He began to travel southward through Galilee—a trip that would ultimately lead Him to Jerusalem and the cross. He desired to travel unnoticed. For the most part, His public ministry was over. Now He wanted to spend time with the disciples, instructing them and preparing them for what lay ahead.

He told them plainly that He was going to be arrested and killed, and that He would rise again after three days. They somehow didn't take it in, and didn't dare to ask Him. "We are often afraid to ask, and in that way lose a blessing."

Contest to be Greatest (9:33-37)

When they reached the house in Capernaum where they would be staying, Jesus asked them what they had been arguing about along the way. They were ashamed to tell Him. They had been disputing as to which of them was the greatest. Perhaps the Transfiguration had revived their hopes for an imminent kingdom, and they were grooming themselves for places of honor in it.

It is really shocking and heartbreaking to realize that at the very time Jesus had been telling them about His impending death, they were esteeming themselves better than others. The human heart is deceitful and desperately wicked above all things, as Jeremiah said.

Jesus knew what they had been talking about so He gave them a lesson on humility. He told them that the way to greatness was by lowly service. The surest way to be first was to voluntarily take the lowest place and live for others instead of self. A little child was set before them and embraced by the Lord Jesus. He emphasized that a kindness shown in His Name to the least esteemed, to the least respected, to the least renowned was an act of greatness. It was the same as if the kindness were shown to the Lord Himself, yes, even to God the Father.

> **The way to greatness is by lowly service.**

"O blessed Lord Jesus, your teachings probe and expose this carnal heart of mine. Break me of self and let your life be lived through me."

Rivalry in Service (9:38-42)

This chapter seems to be full of failures:

- Peter spoke clumsily on the Mount of Transfiguration (v. 5).
- The disciples were defeated in their efforts to cast out the dumb demon (v. 18)
- The disciples strove among themselves as to who was greatest (v. 34).
- Now they demonstrate a sectarian spirit (vv. 38-40).

It was John the beloved who reported to Jesus that they had found a man casting out demons in His Name. The disciples told him to stop because he didn't identify himself with them. It wasn't that the man was teaching false doctrine or living in sin. It was simply that he did not join up with the disciples in their way of things.

> They drew a circle that shut me out–
> Rebel, heretic, thing to flout;
> But love and I had the wit to win–
> We drew a circle that took them in.

Jesus said in effect, "Don't stop him. If he has enough faith in Me to use My Name in casting out demons, he is on My side and is working against Satan. Anyone who has such confidence in the power of My Name isn't apt to turn around quickly and speak evil of Me."

Verse 40 seems to contradict Matthew 12:30 where Jesus said: "Whoever is not with me is against me, and whoever does not gather with me scatters." But there is no real conflict. In Matthew's Gospel, the issue was whether Christ was indeed the Son of God or a demon-empowered man. With such a fundamental question at stake, anyone who is not with Him is against Him, and whoever does not work with Him works against Him.

In the passage in Mark, there was no question as to the Person or work of Christ. It was simply a matter of one's associates in the service of the Lord. Here there must be tolerance and love. Whoever is not against Him in service must be against Satan and therefore on Christ's side.

Even the smallest kindness done in the Name of Christ will be rewarded. A cup of cold water given to a disciple because he belongs to Christ will not go unnoticed. Casting out a demon in His Name is rather spectacular. Giving a glass of water is commonplace. But both are precious to Him when done for His glory.

> **We must consider what effect our words and actions will have on others.**

"Because you belong to Christ" (v. 41). This is the cord that should bind believers together. If we kept these words before us, they would deliver us from party spirit, from petty bickering and from jealousy in Christian service.

Constantly the Lord's servant must consider what effect his words and actions will have on others. It is possible to stumble a fellow believer, and in so doing to cause life-long spiritual damage. It would be better to be drowned with a great millstone about one's neck than to cause a little one to stray from the path of holiness and truth.

There were two kinds of millstones in New Testament times. One was turned by hand. The other was so big it required a donkey to turn it. It is the latter—the great millstone—that Jesus referred to here.

Ruthless Self-Discipline (9:43-50)

The remaining verses of the chapter emphasize the necessity of discipline and renunciation. Those who set out on the path of true discipleship must constantly battle with natural desires and appetites. To cater to them spells ruin. To control them insures spiritual victory.

The Lord spoke of three physical organs—the hand, the foot, and the eye. He explained that it would be better to lose one of these than to be stumbled by it into hell. Reaching the goal is worth any sacrifice.

The hand might suggest our deeds, the foot our walk, and the eye the things we crave after. All these are potential danger spots. Unless these are dealt with severely, they can lead to eternal ruin.

Does this passage teach that true believers can finally be lost and spend eternity in hell? Taken by itself it might suggest that. But taken with the consistent teaching of the New Testament, we must conclude that anyone who goes to hell was never a genuine Christian at all.

A person might *profess* to be born again and appear to go on well for some time. But if that person consistently indulges the flesh, sacrificing the future for present gratifications, it is clear they were never saved.

The Lord repeatedly speaks of hell as a place of fire that never shall be quenched: Where the worm does not die and the fire is not quenched.

It is tremendously solemn. If we really believed it, we would not live for things but for never-dying souls. "Give me a passion for souls, O Lord!"

Fortunately it is never morally necessary to amputate a hand or foot or to cut out an eye. Jesus did not suggest that we should practice such extreme forms of immolation. All He said was it would be better to sacrifice the use of these organs than to be dragged down to hell by their abuse.

Verses 49 and 50 are especially difficult. Therefore we will examine them clause by clause.

"For everyone will be salted with fire …" Here there are three main problems:

1. What "fire" is referred to?
2. What is meant by "salted"?
3. Does "everyone" refer to saved, to unsaved, or to both?

Fire may mean hell (as in vv. 44, 46, 48) or judgment of any kind, including divine judgment of a believer's works, and self-judgment.

Seasoned, translated "salt" in some versions, is used as a figure of that which preserves, purifies, counteracts corruption and seasons. In eastern lands, it is also a pledge of loyalty, friendship or faithfulness to a promise.

> "Give me a passion for souls, O Lord!"

If "everyone" means the unsaved, then the thought is that they will be preserved in the fires of hell, that is, that they will suffer eternal punishment.

If "everyone" refers to believers, the passage teaches that they must:

1. Be purified through the fires of God's chastening in this life, or
2. Preserve themselves from corruption by practicing self-discipline and self-renunciation.

"And every sacrifice will be seasoned with salt" (Mark 9:49 NKJV). This exact clause is omitted in most of the later Bible versions. It is quoted from Leviticus 2:13 (see also Numbers 18:19; 2 Chronicles 13:5). Salt was an emblem of the covenant between God and His people. The salt was intended to remind the people that the covenant was a solemn treaty to be kept inviolate.

In presenting our bodies as a living sacrifice to God (Romans 12:1-2), we should season the sacrifice with salt, that is, we should make it an irrevocable commitment. No turning back!

"Salt is good ..." Christians are the salt of the earth (Matthew 5:13). God expects them to exert a healthful, purifying influence. As long as they fulfill their discipleship, they are a blessing to all.

"But if the salt has lost its saltiness, how will you make it salty again?" Salt without its saltness is valueless. And a Christian who is not carrying out his duties as a true disciple is barren and ineffective. It is not enough to make a good start in the Christian life. There must be constant and radical self-judgment; otherwise the child of God is failing to achieve the purpose for which God saved Him.

"Have salt in yourselves." Be a power for God in the world. Exert a beneficial influence for the glory of Christ. Be intolerant of anything in your life that might lessen your effectiveness for Him.

"And be at peace with one another." This apparently refers back to verses 33 and 34 where the disciples had argued over which of them was the greatest. Pride must be put away and replaced by humble service for all.

To summarize, verses 49 and 50 seem to picture the believer's life as a sacrifice to God. It is salted with fire, that is, mixed with self-judgment and self-renunciation. It is salted with salt, that is, offered with a pledge of unalterable devotedness. If the believer goes back on his vows, or fails to deal drastically with sinful desires, then his life will be savorless, worthless, and pointless. Therefore he should eradicate anything from his life that would interfere with his divinely-appointed mission, and he should maintain peaceful relations with other believers.

LESSON 7 EXAM

Use the exam sheet at the back of the course to complete your exam.

1. **When the Lord Jesus said that some of the disciples would not taste of death till they saw the kingdom of God coming with power, probably**
   A. He was mistaken since all the apostles are long since dead and the kingdom of God has not come.
   B. He was referring to the spiritual kingdom He would establish in the hearts of men.
   C. He was referring to Peter, James and John and the transfiguration that was a foretaste of His coming reign.
   D. He was implying that some of the apostles would not die but would be caught up alive into heaven.

2. **The conditions that prevailed on the Mount of Transfiguration**
   A. foreshadow the millennial reign of Christ.
   B. were normal for that time of the year in that particular place.
   C. reflect the glory of Israel's golden age under Solomon.
   D. illustrate what happens in the human heart when the Lord Jesus has His rightful place.

3. **After the dazzling experience on the mount, the disciples were instructed by the Lord Jesus**
   A. never to disclose what had happened since people simply would not believe it.
   B. to tell their experience only after the Lord had risen from the dead.
   C. to tell the other disciples only.
   D. to broadcast to the whole nation that the King of Israel had come and should be crowned at once.

4. **When confronted with the case of demon possession at the foot of the Mount, the Lord Jesus**
    A. chided His disciples for their unbelief and inability to cast out the evil spirit.
    B. sympathized with His disciples for their powerlessness since this was a particularly difficult case.
    C. gave the disciples a fresh endowment of power to save face before the people and cast out the demon.
    D. chided them, and gave them a second chance to cast out the demon.

5. **When the father of the demon-possessed boy asked the Lord to do something if He could,**
    A. the Lord immediately responded to the appeal by casting out the demon.
    B. the Lord told him that it was not a question of His ability to heal, but rather of the father's ability to believe.
    C. the Lord overlooked the flaw in the man's faith but afterwards drew it to the disciples' attention.
    D. the Lord said He could do no mighty work there because of the man's unbelief.

6. **To teach the disciples their error in seeking positions of worldly advantage in His kingdom, the Lord Jesus**
    A. deliberately took the low seat at the supper table that night.
    B. made them wash one another's feet.
    C. told them that the highest posts of honor were already taken by Abraham, Moses and Elijah.
    D. set a child before them as an object lesson in humility.

7. **When the disciples found a man, who was not one of them, casting out demons in Jesus' name**
    A. they encouraged him
    B. they helped him
    C. they told him to stop
    D. they praised him

8. **Jesus' statement "he who is not against us is on our side" as to do with or is a question of**
   A. the fundamental issue of who He is.
   B. one's associates in the Lord's service.
   C. both doctrine and association.
   D. None of the above.

9. **The Lord's words about cutting off the hand or the foot or plucking out the eye teach**
   A. that true believers, unwilling to take drastic actions in self discipline, may lose their salvation.
   B. that deliberately maiming oneself to gain victory over evil habits is pleasing to the Lord.
   C. that our deeds, our walk and the things which we crave need to be dealt with severely.
   D. that such actions as penance or punishing ourselves promote spirituality.

10. **From the Lord's words about salting the sacrifice with salt and fire we conclude that**
    A. all Christians must be purged by the fires of Purgatory.
    B. only those who are burned at the stake meet the high standard of devotion required.
    C. those who fail to produce this kind of life will perish eternally.
    D. the believer's life of sacrifice must be characterized by self-judgment and devotedness, otherwise it is worthless.

What Do You Say?

On the day Adam and Eve sinned, what promise did God make and what did He do that shows His justice, mercy, and grace?

High Standards

Mark 10

Teaching on Divorce (10:1-12)

From Galilee, our Lord traveled southeastward to Perea, the district on the east side of the Jordan. His Perean ministry extends through verse 45 of this chapter.

It was not long before the Pharisees found Him. They were moving in for the kill, like a pack of wolves. In an effort to trap Him, they asked Him if divorce was lawful. He referred them back to the Pentateuch. What did Moses *command*?

They avoided His question by stating what Moses *permitted*. He allowed a man to put away his wife, provided he gave a written bill of divorce to her. But that was not God's ideal; it was permitted only because of the hardness of the people's hearts. The divine plan is for a man and woman to be joined in marriage as long as they live. This goes back to creation when God made the sexes. A man is expected to leave his parents and be so united in marriage that he and his wife are one flesh. Thus joined by God, they should not be separated by human decree.

Apparently this was difficult for even the disciples to accept. At that time, women did not have a place of honor or security. They were often treated with little more than contempt. A man could put away his wife if he was displeased with her. The wife had no recourse. In many cases, she might as well have been a piece of property.

When the disciples questioned the Lord further, He said pointedly that remarriage after divorce was adultery, whether the man or the woman got the divorce.

If this verse were taken by itself it would be clear that divorce is forbidden under any and all circumstances. But in Matthew 19:9, He made an exception. Where one partner has been guilty of fornication, the other is permitted to get a divorce and is presumably free to remarry.

The Christian faith gives a place of honor to women.

It is also possible that 1 Corinthians 7:15 permits divorce when an unbelieving husband deserts a Christian wife.

It must be confessed that there are difficulties connected with the whole subject of divorce and remarriage. People can get themselves into marital tangles so involved that it takes the wisdom of a Solomon to extricate them. The best way to avoid these tangles is to avoid divorce. Wherever there is divorce there is a cloud and a question mark over the lives of the people involved.

When divorced persons seek fellowship in a local church, the elders must review the case in the fear of God. Every case is different and must be considered individually.

Blessing on Little Children (10:13-16)

The previous paragraph shows Christ's concern for the sanctity of the marriage relationship and for the rights of the women especially. The Christian faith gives a place of honor to women that is not found in other religions or cultures.

Now we see the solicitude of the Lord Jesus for little children. Parents who brought their children to be blessed by the Teacher Shepherd were shooed away by the disciples.

> But Jesus saw them ere they fled,
> And sweetly looked, and gently said,
> "Suffer the children to come unto Me."

Jesus explained that the kingdom of God belongs to little children, and to those who have childlike faith and humility. Adults have to become like little children in order to enter the kingdom.

Certainly these verses should impress the servant of the Lord with the importance of reaching little ones with the Word of God. The minds of children are most plastic and most receptive. "Be your best and give your best to the children" (W. Graham Scroggie).

Sell All (10:17-22)

A rich man intercepted the Lord with what appears to be a sincere inquiry. Addressing Jesus as "Good Teacher," he asked what he had to do to inherit eternal life.

Jesus seized on the words "Good Teacher." He did not refuse the title but used it to test the rich man's faith. Only God is good. Was the rich man willing to confess the Lord Jesus as God? Apparently he was not.

Next the Savior used the law to produce the knowledge of sin. The man was still under the delusion that he could inherit the kingdom on the principle of *doing*. Then let him obey the law. The law told him what to *do*. Our Lord quoted the five commandments that deal primarily with our relations to our fellow man. These five commandments say, in effect, "You shall love your neighbor as yourself." The man professed to have kept them from his youth.

But did he really love his neighbor as himself? If so, let him prove it by selling all his property and giving the money to the poor. Oh, this was another story. He went away sad because he had great possessions.

The Lord Jesus did not mean that this man could have been saved by selling his possessions and giving the proceeds to charity. There is only one way of salvation—that is, by faith in the Lord.

But in order to be saved a person must acknowledge that they are a sinner and that they have fallen short of God's holy requirements. The Lord took the man back to the Ten Commandments to produce conviction of sin. But the rich man's unwillingness to share his possessions showed that he did not love his neighbor as himself. He should have said, "Lord, if that's what is required, then I'm a sinner. I cannot save myself by my own efforts. Therefore I ask you to save me by your grace." But he loved his property too much. He was unwilling to give it up. He refused to break.

> There is only one way of salvation —that is, by faith in the Lord.

We repeat that when Jesus told the man to sell all, He was not giving this as the way of salvation. He was showing the man that he had broken the law of God and that he therefore needed to be saved. If he had responded to the Savior's instruction, he would have been given the way of salvation.

But there is a problem here. Are we who are believers supposed to love our neighbor as ourselves? Does Jesus say to us, "Sell all that you have and give to the poor, and you will have treasure in heaven; and come, follow me"? Each one must answer for himself, but before doing so, he should consider the following inescapable facts:

- Thousands of people die daily of starvation.
- More than half the world has never heard the good news.
- Our material possessions can be used now in the alleviation of human need, both spiritual and physical.
- The example of Christ teaches us that we should become poor that others might be made rich (2 Corinthians 8:9).
- The shortness of life and the imminence of the coming of the Lord teach us to put our money to work for Him now. After He comes it will be too late.

The Peril of Riches (10:23-31)

As He saw the rich man fade into the crowd, Jesus remarked on the difficulty of rich people entering into the kingdom of God. The disciples were amazed by this remark; they linked riches with the blessing of God. "It is easier," He continued, "for a camel to go through the eye of a needle than for a rich person to enter the kingdom of God."

This made the disciples wonder who can be saved. They must have felt that everyone was either rich or wished he were. Jesus answered that what is humanly impossible is divinely possible.

What are we to conclude from the teaching of this passage?

1. First of all, it is especially difficult for rich people to be saved (v. 23). The reason is that these people tend to love their wealth more than they love God. They would rather give up God than give up their money. They put their trust in riches rather than in the Lord. As long as these conditions exist, they cannot be saved.

2. The disciples, being Jewish, had always looked upon riches as a sign of God's favor. This was true in the Old Testament. But now all that is changed. Instead of a mark of the Lord's blessing, riches are a test of a man's devotedness.

3. A camel can go through a needle's eye more easily than a rich man can go through the door of the kingdom. The needle's eye has often been explained as the small door in a city gate. A camel could get through it only by crouching down and hunching along. It was a difficult process.

- But it seems clear from the following verses that the Lord was no longer speaking of difficulty but impossibility. Humanly speaking, a rich man simply cannot be saved.
- Someone may object here that humanly speaking, no one can be saved. That is true. But it is even more true in the case of a rich man. He faces obstacles that the poor man isn't aware of. The god of mammon must be torn from the throne of his heart, and he must stand before God as a pauper. To effect this change is humanly impossible. Only God can do it.
4. Christians who lay up treasure on earth generally pay for their disobedience in the lives of their children. Very few children from such families go on well for the Lord.

Peter caught the drift of the Savior's teaching (v. 28). He realized that Jesus was saying, "Forsake all and follow Me." Peter was right. That is exactly what our Lord intended. Jesus confirmed this by promising present and eternal reward to those who forsake all for His sake and the Gospel's.

1. The present reward is 10,000 percent return, not in money, but in:
   - houses—that is, homes of other people where he is given accommodations as a servant of the Lord.
   - brethren and sisters and mothers—Christian friends whose fellowship enriches all of life.
   - lands—countries of the world that he has claimed for the King.
   - persecutions—these are a part of the present reward. It is a cause of rejoicing when one is found worthy to suffer for Jesus' sake.
2. The future reward is eternal life. This does not mean that we earn eternal life through forsaking all. Eternal life is a gift. Here the thought is that those who forsake all are rewarded with a greater capacity for enjoying eternal life in heaven. All believers will have that life but not all will enjoy it to the same extent.

Then our Lord added a word of warning, "Many who are first will be last, and the last first." It isn't enough to start out well on the path of discipleship. It's how we finish that counts.

"Not everyone who gave promise of being a faithful and devoted follower would continue in the path of self-denial for Christ's Name's sake, and some who seemed backward and whose devotedness was questionable would prove real and self-effacing in the hour of trial" (H. A. Ironside).

Third Prediction of Death and Resurrection (10:32-34)

The time had now come to go to Jerusalem. For the Lord Jesus this meant the sorrow and suffering of Gethsemane, the shame and agony of the cross. What were the emotions of the Son of God at such a time? Can we not read them in the words: "Jesus was walking ahead of them."

- There was determination to do the will of God, knowing full well what the cost would be. He set His face as a flint to go to the city of His death.
- There was loneliness. He was out ahead of the disciples—walking alone. "No one knows but he who has endured it, the solitude of a soul that has outstripped its fellows in zeal for the Lord of hosts. It dare not reveal itself, lest men count it mad. It cannot conceal itself, for a fire burns within its bones. Only before the Lord does it find rest" (C. H. Spurgeon).
- There was joy—a deep, settled joy of being in the Father's will, a joyful prospect of coming glory, the joy of redeeming a bride to Himself. For the joy that was set before Him, He endured the cross, despising the shame.

As we gaze upon Him, striding in the vanguard, we too are amazed. Our intrepid Leader, the Author and Finisher of faith, our glorious Master, Prince divine.

"Let us pause to gaze on that face and form, the Son of God, going with unfaltering step toward the cross! Does it not awaken us to new heroism, as we follow; does it not awaken new love as we see how voluntary was His death for us; yet do we not wonder at the meaning and the mystery of that death?" (Erdman).

Those who followed were afraid. They knew that the religious leaders in Jerusalem wanted His death.

For the third time Jesus gave His disciples a detailed account of coming events. This prophetic outline shows Him to be more than a mere man:

- They are to go up to Jerusalem (11:1-13:37).
- The Son of Man will be delivered to the chief priests and to the scribes (14:1-2; 43-53).
- They will condemn Him to death (14:55-65).
- And will deliver Him to the Gentiles (15:1).

- They will mock Him, and will scourge Him, and will spit on Him, and will kill Him (15:2-38).
- And the third day He will rise again (16:1-11).

Places of Honor in the Kingdom (10:35-45)

Following this poignant prediction of His approaching crucifixion, James and John came with a request that was at once noble and ill-timed, that showed faith and self-seeking. It was noble in that they wanted to be near Christ, but it was a poor time to be seeking great things for themselves. They exhibited faith that Jesus would set up His kingdom, but they should have been thinking of His impending passion.

Jesus asked them if they were able to drink of His cup and share His baptism. His cup, of course, refers to His suffering and His baptism to His death. They professed to be able, and He said they were right. They would suffer because of their loyalty to Him, and James at least would be martyred (Acts 12:2).

Greatness in Christ's kingdom is marked by service.

But then He explained that positions of honor in the kingdom were not bestowed arbitrarily. They would be earned. The measure of a believer's suffering would determine the extent of his honor.

It is good to remind ourselves here that admission to the kingdom is by grace through faith, but position in the kingdom will be determined by faithfulness to Christ.

The other ten disciples were indignant that James and John would try to get ahead of them. But their indignation betrayed the fact that they had the same spirit. This provided the occasion for the Lord Jesus to give a most beautiful and most revolutionary lesson on greatness. Among the unconverted, great people are those who rule with arbitrary power, who are overbearing, who are domineering. But greatness in Christ's kingdom is marked by service. The one who wishes to be first should become a bondslave to everyone.

The Supreme Example is the Lord Jesus Himself. He came not to be served, but to serve and to give His life a ransom for many. Think of it!

- His miraculous birth—He came
- His wonderful life—He ministered
- His vicarious death—He gave His life

As mentioned before, verse 45 is the key verse of the entire Gospel. It is a theology in miniature, a vignette of the greatest Life the world has ever known.

Healing of Blind Bartimaeus (10:46-52)

The scene now shifts from Perea to Judea. The Lord and His disciples had crossed the Jordan and had come to Jericho.

There He met blind Bartimaeus, a man with:

- A desperate need
- A knowledge of the need
- A determination to have it met

The blind man recognized our Lord as the Son of David, and addressed Him as such. It was an odd circumstance. While the nation of Israel was blind to the presence of the Messiah, a blind Jewish man had true spiritual sight.

His persistent pleas for mercy did not go unanswered. His specific prayer for sight brought a specific answer. His gratitude was expressed in faithful discipleship. He followed the Lord on His last trip to Jerusalem. It must have cheered the heart of Jesus to find faith like this in Jericho as He moved on toward the cross.

LESSON 8 EXAM

Use the exam sheet at the back of the course to complete your exam.

1. **The Lord's teaching on divorce**
   A. was based on what was permitted by Moses in the Old Testament law.
   B. was based on the original ideal when God made the sexes.
   C. was based on the customs of His day.
   D. was based on rabbinical traditions.

2. **According to this lesson**
   A. the evangelization of adults is more important than the evangelization of children.
   B. children are of supreme importance and adults must become like little children to enter the kingdom.
   C. children should be seen and not heard.
   D. children do not need to be evangelized since they are already in the kingdom of God.

3. **The rich man who came wanting to do something to inherit eternal life**
   A. was taught by the Lord that he could be saved by giving away his wealth.
   B. was told by the Lord he must give his wealth to Him.
   C. was confronted by the Lord with the hidden sinfulness of his heart.
   D. gladly accepted the Lord Jesus as his personal Savior from sin.

4. **The rich man wanting to do something to inherit eternal life**
   A. responded to the Lord's instructions.
   B. went away sorrowful.
   C. lived up to his claim that he had always kept the commandments.
   D. was willing to acknowledge Jesus to be God.

5. **According to the Lord Jesus, it is especially difficult for**
   A. a child to be saved.
   B. a Gentile to be saved.
   C. an old man to be saved.
   D. a rich man to be saved.

6. **Those who forsake all to follow Christ**
    A. are guaranteed an easy life.
    B. can be more sure of their salvation than believers who do not.
    C. will receive both a present and an eternal reward.
    D. are the only ones who will be ultimately saved.

7. **Comparing the announcement of His death by the Lord in Mark 10:32-34 with two previous announcements (Mark 8:31; 9:31) we observe that**
    A. the three accounts are contradictory.
    B. the Lord only dimly apprehended what lay ahead of Him in Jerusalem.
    C. the third account is more detailed than either of the others.
    D. all three accounts contain identical details.

8. **James and John's request to be allowed to sit on either side of the Lord in His kingdom**
    A. showed how little they understood the Lord's statement about going to the cross.
    B. was self-seeking and brought about a stinging rebuke from the Lord.
    C. showed a real grasp of kingdom truth and understanding of the way high office in the kingdom will be apportioned.
    D. was roundly applauded by the other ten disciples.

9. **The key verse in the Gospel of Mark says**
    A. "the Son of Man hath power on earth to forgive sins."
    B. "a prophet is not without honor, but in his own country."
    C. "He has done all things well."
    D. "For even the Son of Man came not to be served but to serve, and to give his life as a ransom for many."

10. **Blind Bartimaeus**
    A. met the Lord in Perea, just as He was about to cross the Jordan for the last time.
    B. called upon the Lord, as the Son of God, to heal him.
    C. had true spiritual insight to see the Lord Jesus for Who He really was.
    D. was frightened by those who told him to keep quiet when he voiced his need.

What Do You Say?

What three things did Bartimaeus have which we need to have if we are to receive spiritual sight?

LESSON 9

The Last Week Begins

Mark 11, 12

Triumphal Entry (11:1-11)

The record of the last week begins here. Jesus had paused on the east slope of the Mount of Olives, near Bethphage (house of unripe figs) and Bethany (house of dates).

The time had arrived for Him to present Himself openly to the Jewish people as their Messiah-King. He would do this in fulfillment of the prophecy of Zechariah (9:9), riding upon a colt. So He sent two of the disciples from Bethany into Bethphage. With perfect knowledge and complete authority, He told them to bring an unbroken colt that they would find tethered. If anyone challenged them they were to say, "The Lord has need of it and will send it back here immediately." The omniscience of the Lord, as seen here, has prompted someone to say, "This is not the Christ of modernism, but of history and of Heaven."

> **For a moment, at least, He was acknowledged as King.**

Everything happened as Jesus had predicted. The colt was found tied at a main intersection in the village. When challenged, the disciples replied as Jesus had told them. Then the people let them go.

Though the colt had never been ridden before, it did not balk at carrying its Creator into Jerusalem. The Lord rode to the city on a carpet of clothing and palm leaves, with the acclamation of the people ringing in His ears. For a moment, at least, He was acknowledged as King.

The people shouted:

- *Hosanna!*—which originally meant "Save, we pray" but which later became an exclamation of praise. Perhaps the people meant, "Save us, we pray, from our Roman oppressors."
- *Blessed is He who comes in the Name of the LORD!*—a clear recognition that Jesus was the promised Messiah (Psalm 118:26).
- *Blessed is the kingdom of our father David that comes in the name of the LORD!*—they thought that the kingdom was about to be set up, with Christ sitting on the throne of David.
- *Hosanna in the highest!*—a call to praise the Lord in the highest heavens, or for Him to save from the highest heavens.

Once in Jerusalem, Jesus went to the temple—not inside the temple but into the courtyard. Presumably it was the house of God, but He was not at home in this temple because the priests and people refused to give Him His rightful place. After looking around briefly, the Savior withdrew to Bethany with the twelve disciples. It was Sunday evening.

Cursing of the Fig Tree (11:12-14)

This incident is the Savior's interpretation of the screaming, tumultuous welcome He had just received in Jerusalem. He saw the nation of Israel as a barren fig tree—it had leaves of profession but no fruit. The cry of Hosanna would soon turn into the blood-curdling cry, "Crucify Him."

There is an obvious difficulty in the account of the cursing of the fig tree. The difficulty is this: He condemned the fig tree because it had no fruit, yet the record distinctly says, "it was not the season for figs." This seems to picture the Savior as unreasonable, temperamental and petulant. We know that this is not true; yet how can we explain this curious circumstance?

Fig trees in Bible lands produced an early fruit before the leaves appeared. This fruit was edible. It was a harbinger of the regular crop, here described as the time of gathering figs. If no early figs appeared, then that was a sign that there would be no regular crop later on.

When Jesus came to the nation of Israel, there were leaves, which speak of profession, but there was no fruit for God. There was promise without fulfillment, profession without reality. Jesus was hungry for fruit from the nation.

Because there was no early fruit, He knew that there would be no later fruit from that unbelieving people, and so He cursed the fig tree.

This pre-pictured the judgment that was about to fall on the nation of Israel in AD 70.

However, the incident does not teach that Israel was cursed to perpetual barrenness. The Jewish people have been set aside temporarily, but when Christ returns to reign, the nation will be reborn and will be restored to a position of favor with God.

This is the only miracle in which Christ cursed rather than blessed. Only here did He destroy life rather than restore it. This has been raised as a difficulty. However, the objection is not valid. The Creator has the sovereign right to destroy an inanimate object in order to teach an important spiritual lesson and thus save people from eternal doom.

Although the primary interpretation of this passage relates to the nation of Israel, it has application to people of all ages who combine high talk and low walk.

Cleansing of the Temple (11:15-19)

At the outset of His public ministry, Jesus had driven merchants out of the temple grounds (John 2:13-22). Now as His ministry drew to a close, He again entered the court of the temple and drove out those who were making profit from sacred activities. He even stopped the carrying of ordinary utensils through the area.

Combining quotations from Isaiah and Jeremiah, He condemned those who were desecrating the temple and using it as a means for personal gain. God had intended the temple to be a house of prayer for all nations (Isa. 56:7), not just for Israel. These men had made it a religious market, a hang-out for con-men and racketeers (Jer. 7:11).

The chief priests and scribes were cut deeply by His accusations. They wanted to kill Him, but they could not do it brazenly because the common people still looked on Him with a great deal of awe and wonder.

Verse 19 mentions that Jesus left Jerusalem every evening. It was His custom to do so, perhaps for safety. He was not afraid for Himself. Part of His ministry was to preserve His sheep, that is, His own disciples (John 17:6-19).

The Fig Tree Withered (11:20-26)

From verse 20 to 13:37, we have another typical day in the life of the Lord Jesus. The first one we noticed was in 1:21-38.

On the morning following the cursing of the fig tree, the disciples passed it on their way to Jerusalem. It had withered away from the roots up. When Peter mentioned this to the Lord, He simply said, "Have faith in God."

But what do these words have to do with the fig tree? How are the Lord's words connected with what Peter had said?

The following verses make it clear that Jesus was encouraging faith as the means to remove difficulties. If disciples have faith in God, they can deal with the problem of fruitlessness, they can remove mountainous obstacles.

However, these verses do not give a person authority to pray for miraculous powers for their own convenience or to attract attention to themselves. Every act of faith must have the promise of God to rest on. If we know that it is God's will to remove a certain difficulty, then we can pray with utter confidence that it will be done. In fact, we can pray with confidence on any subject as long as we are confident it is according to God's will as revealed in the Bible or by the inner witness of the Spirit.

When we are really living in touch with the Lord and praying in the Spirit, it is possible to have the assurance of answered prayer before the answer actually comes.

But one of the basic requirements for answered prayer is a forgiving spirit. If we nurse a harsh, vindictive attitude toward others, we cannot expect God to hear and answer us. We must forgive if we are to be forgiven. This does not refer to the judicial forgiveness of sins at the time of conversion; that is strictly a matter of grace through faith. But this refers to God's parental dealings with His children. An unforgiving spirit in a believer breaks fellowship with God and hinders the flow of blessing.

Christ's Authority Questioned (11:27-33)

As soon as He reached the temple area, the religious leaders accosted Jesus and challenged His authority by asking two questions:

1. By what authority are you acting as you do?
2. Who gave you the authority to do these things, that is, to cleanse the temple, to curse the fig tree, and to ride triumphantly into Jerusalem?

They hoped to trap Him, no matter how He answered. If He had claimed to have authority in Himself as the Son of God, they would have accused Him of blasphemy. If he had claimed authority from men, they would have discredited Him. If He had claimed to have received authority from God, they would have challenged the claim; they considered themselves the God—appointed religious leaders of the people.

Jesus answered them by asking a question. Was John the Baptist divinely commissioned or not? (The expression "the baptism of John" refers to his entire ministry.) They couldn't answer without embarrassment. If John's ministry was divinely appointed, they should have obeyed his call to repent. If they disparaged John's ministry, they would risk the anger of the common people who still considered John a spokesman for God.

When they refused to answer, professing ignorance, the Lord refused to discuss His authority. As long as they were unwilling to acknowledge the credentials of the forerunner, it was obvious they would not acknowledge the higher credentials of the King Himself.

The Wicked Tenants (12:1-12)

The Lord Jesus was not through with the Jewish authorities, even if He had refused to answer their question. He now delivered a stinging indictment of them for their rejection of God's Son. The indictment was in the form of a parable.

The man who planted a vineyard was God Himself. The vineyard was the place of privilege that was then occupied by Israel. The hedge was the law of Moses that separated Israel from the Gentiles and preserved them as a distinct people for the Lord. The tenants were the religious leaders, such as the Pharisees, the scribes, and the elders.

Repeatedly God sent His servants, the prophets, to the people of Israel, seeking for fellowship, for holiness, for love. But the people persecuted the prophets and killed some of them.

Finally God sent His beloved Son. Surely they would respect Him. But they didn't. They plotted against Him and finally killed Him. Thus the Lord predicted His own death and exposed His guilty murderers.

> One of the basic requirements for answered prayer is a forgiving spirit.

What would God do with such wicked men? He would utterly destroy them and give the place of privilege to others. The "others" here may refer to the Gentiles, or to the repentant remnant of Israel in the last days.

All this was in fulfillment of the Old Testament Scriptures. In Psalm 118:22, for example, it was prophesied that the Messiah would be rejected by the Jewish leaders in their building plans. They would have no place for this Stone. But following His death, He would be raised from the dead and given the place of preeminence by God. He would be made the topmost stone in God's building.

The Jewish leaders got the point. They believed that Psalm 118 spoke of the Messiah. Now they heard the Lord Jesus applying it to Himself. They wanted to seize Him and destroy Him, but the time had not come. The crowd would have taken sides with Jesus. So they left Him for the time being.

Caesar or God (12:13-17)

This chapter contains attacks on the Lord by the Pharisees and Herodians and by the Sadducees. It is a chapter of questions. (See vv. 9, 10, 14, 15, 16, 23, 26, 28, 35, 37.)

The Christian should always maintain a good testimony before the world.

The Pharisees and Herodians were bitter foes, but now they were brought together by a common hatred of the Savior. They desperately tried to trick Him into saying something that they could use as a charge against Him. So they asked if it was lawful to pay taxes to the Roman government.

No Jewish person particularly enjoyed living under Gentile rule. The Pharisees hated it with a passion, whereas the Herodians adopted a more tolerant view. If Jesus openly endorsed paying tribute to Caesar, He would alienate many of the Jewish people. If He spoke against Caesar, they would hustle Him to the Roman authorities for arrest and trial as a traitor.

Jesus asked someone to bring a denarius. (Apparently He Himself did not have one.) The coin bore the image of Tiberius Caesar, a reminder to the Jewish people that they were a conquered, subject people. Why were they in this condition? Because of their unfaithfulness and sin. They should have been humbled and ashamed to have to admit that the coins they used had the image of a Gentile dictator on them.

Jesus said to them, "Render to Caesar the things that are Caesar's, and to God the things that are God's." Their great failure had not been in the first area but in the second. They had paid their Roman taxes, though reluctantly, but they had disregarded the claims of God on their lives.

E. A. Abbott has a suggestive thought. The coin had Caesar's image upon it, and therefore belonged to Caesar. Humans have God's image upon them—God created humanity in His own image (Genesis 1:26-27)—and therefore belongs to God.

The believer is obligated to obey and support the government under which he lives. He is not to speak evil of his rulers or work for the overthrow of the government. He is to pay taxes and to pray for those in authority. If he is called upon to do anything that would violate his higher loyalty to Christ, then he is to refuse and to bear the punishment. The claims of God must come first. In upholding those claims, the Christian should always maintain a good testimony before the world.

The Sadducees and the Resurrection (12:18-27)

The Sadducees were the liberals or modernists of that day. They scoffed at the idea of bodily resurrection. So they came to the Lord with a preposterous story, trying to ridicule the whole idea.

They reminded Jesus that the law of Moses made special provision for widows in Israel. In order to preserve the family name and to keep the property in the family, the law stipulated that if a man died childless, his brother should marry the widow (Deuteronomy 25:5–10).

Here was a fantastic case in which a woman married seven brothers, one after the other. Then she died last of all. Now for their clever question! Whose wife would she be in the resurrection?

They thought they were smart; the Savior told them they were extremely ignorant—ignorant of the Word of God which teaches resurrection and ignorant of the power of God which raises the dead.

First they should know that the marriage relationship does not continue in heaven. Believers will recognize one another in heaven and will not lose their distinctions as men and women, but they will not marry nor give in marriage. In that respect, they will resemble the angels.

Then our Lord took the Sadducees back to the books of Moses that they valued above the rest of the Old Testament. Specifically He took them back to the account of Moses at the burning bush (Exodus 3:6). There God spoke of Himself as the God of Abraham, and the God of Isaac, and the God of Jacob. The Savior used this to show that God is the God of the living, not of the dead.

God is the God of the living, not of the dead.

But how so? Weren't Abraham, Isaac, and Jacob dead when God appeared to Moses? Yes, their bodies were in the Cave of Macphelah in Hebron. How then is God the God of the living?

The argument seems to be this:

- God had made promises to the patriarchs concerning the land and concerning the Messiah.
- These promises were not fulfilled during their life times.
- When God spoke to Moses at the burning bush, the bodies of the patriarchs were in the grave.
- Yet God spoke of Himself as the God of the living.
- He must fulfill His promises to Abraham, Isaac, and Jacob.
- Therefore resurrection is an absolute necessity from what we know of the character of God.

And so the Lord's parting word to the Sadducees was, "You are quite wrong."

The Scribe's Question (12:28-34)

One of the scribes was sincerely impressed by our Lord's handling of His critics' questions. So he asked Jesus what is the most important of all the commandments. It was an honest question, and, in some ways, the most basic question in life. What he was really asking for was a concise statement of the chief aim of man's existence. What is really important in life?

Jesus began with a quotation from the Shema, a Jewish statement of faith taken from Deuteronomy 6:4: "Hear, O Israel: The LORD our God, the LORD is one."

Then He summed up man's responsibility to God: Love Him with the entirety of one's heart, soul, mind and strength. God is to have the supreme place in man's life. No other love can be allowed to rival love for God.

The other half of the Ten Commandments teaches us to love our neighbor as ourselves. We are to love God *more* than ourselves, and our neighbor *as* ourselves.

In other words, the life that really counts is the one that is concerned first with God, then with others. Material things are not mentioned. God is important and people are important.

God is concerned with what a person is inwardly.

The scribe agreed heartily with the Lord's answer. He stated with commendable clarity that love to God and love to one's neighbors were far more important than rituals. He realized that people could go through religious ceremonies and put on a great, public display of piety without inward, personal holiness. He acknowledged that God is concerned with what a person is inwardly as well as what they are outwardly.

When Jesus heard this remarkable observation, He told the scribe that he was not far from the kingdom of God. True subjects of the kingdom do not try to deceive God or their fellow-men or themselves with an external religion. Realizing that God looks on the heart, they go to Him for cleansing for sin and for power to live in a manner that is pleasing to Him.

After this, no one tried to trap Jesus by asking Him leading questions.

David's Son and Lord (12:35-37)

The scribes had always taught that when the Messiah came, He would be a lineal descendant of David. That was true but it was not the whole truth. So the Lord Jesus now posed a problem to those gathered around Him in the temple court. In Psalm 110:1, David spoke of the coming Messiah as his Lord. How could this be? How could the Messiah be David's Son and his Lord at the same time?

The answer is clear to us. The Messiah would be both Man and God. As David's Son, He would be human. As David's Lord, He would be divine.

The common people heard Him gladly. Apparently they were willing to accept the fact, even if they might not have understood it fully. But nothing is said of the Pharisees and scribes. Their silence is ominous.

Warning Against Scribes (12:38-40)

The scribes were outwardly religious but inwardly wicked. Notice their outward show;

- They loved to parade in long robes. This was designed to distinguish them from the common people and to give them a sanctimonious appearance.
- They loved to be greeted with high sounding titles in public places. It did something for their ego.
- They sought the places of honor in the synagogue, as if physical location had something to do with godliness.
- They not only wanted religious prominence, but social distinction as well. They wanted the best places at feasts.

Inwardly they were greedy and insincere. They robbed widows of property and livelihood in order to enrich themselves, pretending, of course, that the money was for the Lord. They recited long prayers—great swelling words of vanity—prayers of words alone.

"They loved *peculiarity* (long clothing); *popularity* (salutations); *prominence* (chief seats); *priority* (uppermost rooms); *possessions* (widows' houses); *mock piety* (long prayers)." (From *Daily Notes* of the Scripture Union.)

The Widow's Mite (12:41-44)

In vivid contrast to the selfish greed of the scribes was the devotedness of this widow. They devoured widows' houses; she gave all she had to the Lord.

The whole scene shows the supernatural knowledge of the Lord. He watched the rich people dropping sizable gifts into the chest for the temple treasury. He knew that their giving did not represent a sacrifice. They gave out of their abundance. He also knew that the two mites she gave was all her living. He therefore announced that she gave more than all the rest put together.

As regards monetary value or buying power, she gave very little. But the Lord estimates giving by our motive, our means, and by how much we have left. This is a great encouragement to those who have little in the way of material possessions, but a great desire to give to Him.

It is amazing how we can approve of the widow's action and agree with the Savior's verdict without imitating her example. Yet if we really believe what we say we believe, we would do exactly what she did. Her gift expressed her conviction that all belonged to the Lord and that He was worthy of all, and must have all.

Many Christians today find fault with her for not providing for her future. Did this show a lack of foresight and prudence? But this is the life of faith—plunging all into the work of God now and trusting Him for the future. Did He not promise to provide food and clothing for those who seek first the kingdom of God and His righteousness? (Matthew 6:33).

Radical? Revolutionary? Unless we see that the teachings of Christ were radical and revolutionary, we have missed the emphasis of His ministry.

LESSON 9 EXAM

Use the exam sheet at the back of the course to complete your exam.

1. **The Lord's triumphal ride into Jerusalem**
    A. was deliberately planned by the Lord Himself.
    B. came about as a spontaneous popular movement which took the Lord by surprise.
    C. fulfilled a prophecy of Habakkuk.
    D. took place twice during His ministry on His visits to Jerusalem.

2. **The cursing of the fig tree by the Lord Jesus**
    A. is the second time where the Lord used His power to destroy life rather than restore it.
    B. was a symbolic act signifying the temporary setting aside of Israel because of its spiritual barrenness.
    C. teaches that the Jewish race is cursed to perpetual spiritual unfruitfulness.
    D. was unreasonable, since "the time of figs was not yet," according to Mark.

3. **The Lord's accusation that the Jews had made the Temple "a den of thieves" is a quotation from**
    A. Zechariah 9:9.   C. Jeremiah 7:11.
    B. Isaiah 56:7.   D. Psalm 118:22.

4. **When the Lord said that faith could move mountains, He meant**
    A. that proper faith could literally command a mountain to pluck itself up and cast itself into the sea.
    B. that God will bestow on the undoubting believer miraculous powers regardless of his motives or needs.
    C. it is strictly rhetorical and not literal or symbolic.
    D. that obstacles can be removed according to God's will in answer to believing prayer.

5. **When the Lord's authority was challenged by the religious leaders**
   A. He refused to answer them.
   B. He countered with a question that they were embarrassed to answer.
   C. He replied that this authority was in Himself as the Son of God.
   D. He accused them of being "a generation of vipers."

6. **When the Lord applied the Old Testament Scripture about the stone rejected of the builders, to Himself the religious leaders**
   A. were surprised because they had never regarded that Scripture as Messianic.
   B. rejoiced because it helped them see that Jesus was indeed the Messiah.
   C. trembled in case they unintentionally become guilty of causing the prophecy to be fulfilled.
   D. were angry because He was applying the prophecy to Himself and to them.

7. **The question regarding paying tribute to Caesar was**
   A. disregarded by the Lord.
   B. a catch question to either alienate the Lord with the Jews or embroil Him with the Romans.
   C. asked in all sincerity since the Jews hated paying taxes to a foreign government.
   D. unanswerable so the Lord refused to answer it.

8. **The question the Sadducees asked regarding the resurrection**
   A. was asked in sincerity because, although the Sadducees believed in the doctrine of resurrection, the particular hypothetical case they mentioned had always bothered them.
   B. demonstrated their ignorance of God's power and His Word.
   C. was acknowledged by the Lord Jesus to be a very clever question.
   D. was answered by the Lord by reminding them that He had already raised Jairus' daughter.

9. The Scribe who asked the question, "Which is the most important commandment of all?"
   A. was told by the Lord Jesus that he was not far from the kingdom of God.
   B. was obviously insincere as his response to the Lord's answer indicates.
   C. was prompted to ask this question by the Herodians who were still seeking to trap the Lord in His words.
   D. was told by the Lord that he should keep the commandments since he could quote them so well.

10. The Lord valued very highly the widow's two mites because she gave without reserve to the Lord. this shows that His teachings are
    A. impractical in this materialistic age.
    B. usually to be taken allegorically rather than literally.
    C. both radical and revolutionary.
    D. dispensational, having to do mostly with the ideal conditions of the Millennial Kingdom.

What Do You Say?

Why do you think that the Lord responded with a question to the religious leaders when they asked Him about His authority?

Signs of Christ's Return

Mark 13

Destruction of the Temple Foretold (13:1-2)

As the Lord Jesus was leaving the temple area for the last time before His death, one of the disciples tried to arouse His enthusiasm concerning the magnificence of the temple and the surrounding buildings and walls. The disciples were occupied with the enormous size of the stones and the architectural triumphs involved in erecting them. The Savior pointed out that these things were soon to be destroyed. Not one stone would be left on another when the Roman armies would invade Jerusalem in AD 70. Why be occupied with materialistic things that are only passing shadows?

The Beginning of Travail (13:3-8)

In His discourse on the Mount of Olives, the Lord diverted the disciples' attention to events of greater importance. Some of the prophecies seem to depict the destruction of Jerusalem, AD 70; most of them obviously go beyond that date to the Tribulation period and to the personal return of Christ in power and glory.

The watchwords of the discourse, which apply to believers in every dispensation, are:

- Be on your guard (vv. 9, 23, 33)
- Do not be alarmed (v. 7)
- Endure (v. 13)

- Pray (vv. 18)
- Stay awake (vv. 33, 35, 37)

The discourse was introduced by a question from four of the disciples (Peter, James, John, and Andrew). When would the temple be destroyed, and what sign would precede the prophesied event? The answer of our Lord included the destruction of Herod's temple, then standing but looked beyond it to the destruction of a later temple, which would take place during the Great Tribulation, prior to His second advent.

First, they were to be on their guard lest anyone lead them astray by falsely claiming to be the Messiah. Many false Christs would appear, as seen in the rise of so many cults, each with its own anti-Christ.

Secondly, they should not interpret wars and rumors of wars as a sign of the end times. All through the intervening period there would be international strife. In addition, there would be great cataclysms of nature—earthquakes, famines, and the like. These would be but preliminary birth pangs, ushering in a period of unparalleled travail.

Persecution of Disciples (13:9-13)

Thirdly, the Lord predicted great personal testing for those who would be unflinching in their testimony for Him. They would be put on trial before religious and civil courts.

While this section is applicable to all periods of Christian testimony, it seems to have special reference to the ministry of the 144,000 during the tribulation period. These are Jewish believers who will carry the Gospel of the kingdom to all nations of the earth prior to Christ's coming to reign. Verse 10 should not be used to teach that the Gospel must be preached to all nations before the rapture. It *should be* proclaimed worldwide and perhaps it *will be*, but to say that it *must be* is to state something the Bible doesn't state. No prophecy needs to be fulfilled before Christ's coming for His saints; He may come at any moment.

The Lord promised that persecuted believers on trial for His sake would be given divine help in making their defense. They would not need to prepare their case in advance; perhaps there would not be time. The Holy Spirit would give them exactly the right words. This promise should not be used as an excuse for not preparing sermons or Gospel messages today. But it is a guarantee of supernatural help for crisis times.

Another feature of tribulation days will be widespread betrayal of those who are loyal to the Savior. Family members will serve as spies and informers against believers. A great wave of anti-Christian sentiment will sweep the world. It will take courage to remain true to the Lord Jesus, but those who do endure to the end will be saved. This cannot mean that they will receive eternal salvation because of their endurance; that would be a false gospel. Neither can it mean that faithful believers will be saved from physical death during the tribulation, because we read elsewhere that many will seal their testimony with their blood. What it probably means is that endurance to the end will evidence reality, that is, it will characterize those who are genuinely saved.

The Abomination of Desolation (13:14-23)

At verse 14, we have come to the middle of the tribulation period, the beginning of the great tribulation. We know this by comparing this passage with Daniel 9:27. At that time, a great abominable idol will be set up in the temple in Jerusalem. People will be compelled to worship it or to be slain. True believers will, of course, refuse.

The setting up of this idolatrous image will signal the beginning of great persecution. Those who read and believe the Bible will know that the time has come to flee from Judea. There will not be time to gather up personal belongings. Pregnant women and nursing mothers will be at a distinct disadvantage. If it happens during the winter, that will add further hazards.

Notice verse 19. It will be a time of tribulation greater than anything in the past or in the future. It is the *Great Tribulation*. The Lord Jesus is not speaking here about the general type of tribulation that Christians in every age have encountered. This is a period of trouble that is unique in its intensity.

Notice too that the tribulation is primarily Jewish in character. We read of the temple (v. 14, compare Matthew 24:15) and of Judea (v. 14). It is the time of distress for Jacob (Jeremiah 30:7). The church is not in view here. It will have already been taken to heaven before the day of the Lord begins (1 Thessalonians 4:13-18; 1 Thessalonians 5:1-3).

The vials of God's wrath will be poured out on the world in those days. It will be a time of calamity, chaos and bloodshed. In fact, the slaughter will be so great that God will supernaturally shorten the period of daylight; otherwise no one would survive. But the days will be shortened in order that God's elect remnant of Jewish believers will be preserved.

The great tribulation period will again witness the rise of false messiahs. People will be so desperate they will turn to anyone who promises them safety. But believers will know that Christ will not appear quietly or unheralded. Even if these false Christs perform supernatural wonders (as they will), the elect will not be deceived. They will realize that these miracles are satanically inspired.

Miracles are not necessarily divine. They represent superhuman departures from the known laws of nature but may represent the work of Satan, angels, or demons. The Man of Sin will be given satanic power to perform miracles (2 Thessalonians 2:9).

This is why believers should be on their guard and be forewarned.

The Glorious Appearing (13:24-27)

After the tribulation, there will be startling disturbances in the heavens. Darkness will shroud the earth both by day and by night. The stars will fall from heaven and the powers that are in the heavens (the forces that keep stellar bodies in orbit) will be shaken.

Then an awe-struck world will see the Son of Man returning to the earth, not now as the lowly Nazarene but as the glorious Conqueror. He will come in clouds, escorted by myriads of angelic beings and of glorified saints. It will be a scene of overwhelming power and dazzling splendor. He will dispatch His angels to gather together all His elect, that is, all who have acknowledged Him as Lord and Savior during the tribulation period. From one end of the earth to the other they will come to enjoy the benefits of His wonderful thousand-year reign on earth. His enemies, of course, will be destroyed at the same time.

Lesson on Watchfulness (13:28-37)

The fig tree is a type of the nation of Israel. Jesus taught here that prior to His second advent the fig tree would put forth leaves. In 1948, the independent nation of Israel was formed. Today that nation exerts an influence in world affairs that is all out of proportion to its size. Israel can be said to be putting forth its leaves. There is no fruit as yet; in fact, there will be no fruit until the Messiah returns to a people who are willing to receive Him.

The formation and growth of the nation of Israel tell us that the King is near, even at the doors. If His coming to reign is that near, how much nearer is His coming for the church!

Verse 30 is often understood to mean that all the things prophesied in this chapter would take place while the people of Christ's day were still living. But it cannot mean that because many of the events, especially verses 24-27, simply did not take place at that time. "This generation" may mean "this race." We believe it means, "this Jewish race characterized by unbelief and rejection of the Messiah." The testimony of history is that "this generation" has not passed away. The nation as a whole has not only survived as a distinct people but has continued in its deep-seated animosity toward the Lord Jesus. Jesus predicted that the nation and its national characteristic would continue until His second advent.

The next verse (31) emphasizes the absolute certainty of every prediction of the Savior. The atmospheric heavens and the stellar heavens will pass away. The earth itself

The Son of Man will return as the glorious Conqueror.

will be dissolved. But every word He spoke will come to pass.

Jesus said, "But of that day and hour no one knows, not even the angels in heaven, nor the Son, but only the Father" (v. 32). It is well known that this verse has been used by enemies of the Gospel to prove that Jesus was nothing more than a man with limited knowledge like ourselves.

It has also been used by sincere but misguided believers to demonstrate that Jesus emptied Himself of the attributes of deity when He came into the world as a man.

Neither of these interpretations is true. Jesus was both God and Man. He had all the attributes of deity and all the characteristics of perfect manhood. It is true that His deity was veiled in a body of flesh, but it was there nonetheless. There was never a time when He was not fully God.

How then can it be said of Him that He does not know the time of His second advent?

We believe the key to the answer is found in John 15:15: " … the servant does not know what his master is doing …" As a perfect Servant, it was not given to the Lord Jesus to know the time of His coming. As God He, of course, knows it. But as Servant, it was not given to Him to know it for the purpose of revealing to others.

"It is not a denial of our Lord's divine omniscience, but simply an assertion that in the economy of human redemption it was not for Him 'to know times or seasons that the Father has fixed by his own authority,' Acts 1:7. Jesus knew that He will come again, and often spoke of His second advent, but it did not fall to His office as Son to determine the date of His return, and hence He could hold it up before His followers as the object of constant expectation and desire" (James H. Brookes).

The chapter closes with an exhortation to watchfulness and prayer in view of the Lord's return. The fact that we do not know the appointed time should keep us on the alert.

A similar situation is common in everyday life. A man goes away from home on a long trip. He leaves instructions with his bondservant and tells the watchman also to be on the lookout for his return.

Jesus likened Himself to the traveling man. He may come back at any hour of the night. His people, serving as night watchmen, should not be found sleeping. So He left this word for all His people, "Stay awake."

LESSON 10 EXAM

Use the exam sheet at the back of the course to complete your exam.

1. **The question which prompted the Olivet Discourse was,**
   A. "How can we know the time of Your coming?"
   B. "When will You restore the kingdom?"
   C. "When will these things be, and what will be the sign when all these things are about to be accomplished?"
   D. "What will the time of trouble be like?"

2. **Wars and natural disasters, such as earthquakes and famines**
   A. should be looked upon as signs of the end times.
   B. are but preliminary signs of the end and will characterize the entire intervening period.
   C. will tend to disappear as the age draws to a close.
   D. are characteristic of our day and indicate we are already in the Tribulation.

3. **Before the Lord Jesus can come for His own**
   A. the Gospel must be preached to all nations.
   B. the apostasy of Christendom must be complete.
   C. the antiChrist must appear on earth.
   D. no prophecy remains to be fulfilled.

4. **The expression "he that shall endure unto the end shall be saved" indicates that**
   A. those who fail to endure to the end will lose their salvation.
   B. eternal salvation is the reward of faithfulness even in the face of persecution.
   C. during the Great Tribulation, all faithful believers will be preserved from harm in spite of persecution by the Antichrist.
   D. endurance to the end will evidence reality and characterize all who are genuinely saved.

5. **The "Abomination of Desolation" is**
   A. the United Nations Organization.
   B. a period of terrible persecution.
   C. an idol that will be set up in the Temple at Jerusalem.
   D. a particularly impure and immoral government set up on earth by the Man of Sin.

6. **"The Great Tribulation" refers to**
   A. the persecution of the Christians under Nero and the other "persecuting Caesars."
   B. present day persecutions of Christians.
   C. all persecutions of God's people from the death of Abel to the end of the present age, such persecutions being viewed collectively.
   D. a special period of persecution specifically of the Jews after the rapture of the Church.

7. **The period of the "Great Tribulation" will be characterized by**
   A. the appearance of false Messiahs.
   B. supernatural wonders to deceive mankind.
   C. terrible carnage and slaughter.
   D. all the above

8. **The author's view of the statement "this generation will not pass away, till all these things take place" refers to**
   A. the generation of which the Lord was speaking rather than the generation to which He was speaking.
   B. all things prophesied in the Olivet Discourse would take place while the men of Christ's day were still living.
   C. the Jewish race, characterized by unbelief and rejection of Messiah would not pass away until all took place.
   D. the generation of Antichrist, a generation still future.

9. **The formation, growth and influence of the State of Israel**
   A. is of passing interest and of no prophetic significance since God has no future for Israel.
   B. has no direct connection with the parable of the fig tree since that parable must be interpreted symbolically.
   C. is of no prophetic significance because the State of Israel does not recognize Jesus as Messiah.
   D. is evident proof that the King is near.

10. **The last exhortation in Mark's record of the Olivet Discourse is that we should**
    A. work for the night is coming.
    B. stay awake because we do not know when the Master of the house will come.
    C. wait patiently for the fruit of harvest as a good famer.
    D. war against the rulers of this world's darkness and against wicked spirits in high places.

What Do You Say?

What is involved in being watchful for the Lord's coming?

The Betrayal and Trial of Jesus

Mark 14

Plot Against Christ (14:1-2)

It was now Wednesday of that fateful week. In two days the Passover would be held, ushering in the seven day Feast of Unleavened Bread. The religious leaders were determined to destroy the Lord Jesus, but they didn't want to do it during the religious holidays because many of the people still considered Jesus to be a prophet.

The chief priests and scribes determined not to kill Him at the time of the Passover. Yet it seems that divine providence overruled them, and that the Paschal Lamb of God was killed at that very time (see Matthew 26:2).

Jesus Anointed by Woman (14:3-9)

Someone has entitled this incident "Love in a frame of hate" because it comes between the plotting of the religious hierarchy and that of Judas.

If this is the same incident recorded in John 12:1-8, it occurred six days before the Passover (John 12:1). Mark does not always follow the chronological order. Here his purpose seems to be to contrast the value that this woman put on Christ with the value Judas put on Him (vv. 10-11).

The bread signified Christ's body given, the cup His blood shed.

Simon the leper held a feast in honor of the Savior, perhaps in gratitude for being healed. An unnamed woman ("Mary" in John 12) lavishly anointed the head of the Lord Jesus with some very expensive perfume. Her love for Him was great.

Some of the guests thought this was a tremendous waste. She was too reckless, too wasteful. Why hadn't she sold the perfume and given the money to the poor? Just think what this would do for the poor!

Three hundred denarii was the equivalent of a year's wages (Unger's Bible Dictionary). People still think it is a waste to give a year of one's life to the Lord. How much more a waste is it to give one's whole life to the Lord!

Jesus rebuked their murmuring. She had recognized her golden opportunity to pay this tribute to the Savior. If they were so solicitous for the poor, they would always be able to help them because the poor are always present. But the Lord would soon die and be buried. This woman wanted to show this kindness while she could. She might not be able to care for His body in death, so she would show her love while He was still alive.

The fragrance of that perfume reaches down to our generation. Jesus said that she would be memorialized worldwide. She has been—through the Gospel records.

The Betrayal Plan (14:10-11)

She prized the Savior highly. Judas, by contrast, valued Him very lightly. Though he had lived with the Lord Jesus for at least a year, and had received nothing but kindness from Him, Judas now sneaked off to the chief priests with a definite guarantee to betray the Son of God into their hands. They seized the offer gladly, offering to pay him money for his treachery. All he had to do now was work out the details.

Preparation for the Passover (14:12-16)

Although the exact chronology is not certain, it is probable that we have now come to Thursday of Passover Week.

The disciples did not realize that this would be the fulfillment and climax of all the Passovers that had ever been held. They asked the Lord for directions as to where to hold the Passover. He sent them to Jerusalem with instructions to look for a man carrying a pitcher of water. Usually the women carried the water pots; to see a man doing it was a rarity. This man would lead them to the proper house. They would then ask the owner to show them to a room where the Teacher could eat the Passover with His disciples.

It is wonderful to see the Lord choosing and commanding in this way. He acts as the Sovereign Ruler of humanity and property. And it is also wonderful to see responsive hearts putting themselves and their possessions at His disposal. It is good for us when He has instant, ready access to every room in our lives!

The Betrayal Discussed (14:17-21)

That same evening, the Lord came with His disciples to the upper room that had been prepared. As they reclined and ate, Jesus announced that one of the disciples would be His betrayer. They all recognized the evil propensities of their own natures. With a healthy distrust of self, each asked if he were the culprit. Jesus then disclosed the traitor as the one who dipped the bread with Him into the dish, that is, the one to whom He gave the morsel of bread (v. 20). The Son of Man was going forward to His death as predicted, He said, but the doom of His betrayer would be great. In fact, it would be better if he had never been born.

The Lord's Supper (14:22-26)

After receiving the morsel of bread, Judas went out into the night (John 13:30). Jesus then instituted what we know as the Lord's Supper. Its meaning is beautifully outlined in the three words:

- He took—humanity upon Himself.
- He broke—He was about to be broken on the cross.
- He gave—He gave Himself for us.

The bread signified His body given, the cup His blood shed. By His blood He ratified the new covenant. For Him there would be no more festive joy until He returned to earth to set up His kingdom.

At that point, they sang a hymn—probably a portion of the Great Hallel—Psalms 113-118. Then they went out from Jerusalem, across the Kidron, to the Mount of Olives.

Peter's Self-Confidence (14:27-31)

On the way, the Savior warned the disciples that they would all be ashamed and afraid to be known as His followers in the hours ahead. It would be as Zechariah had predicted; the Shepherd would be smitten and His sheep would flee in panic (Zechariah 13:7). But then He graciously assured them that He would not disown them; after rising from the dead, He would be waiting for them in Galilee.

Peter was indignant at the thought of denying the Lord. The others might do it, but he?—Never! Jesus corrected that "Never" to "Soon." Before the rooster crowed twice, Peter would have disclaimed connection with the Savior three times.

"Preposterous," shouted Peter, "I'll die before I deny You."

Peter wasn't the only one to make that noisy boast. They all engaged in brash, self-confident assertions. And we are no different. Let us never forget that. We are no different. We all have to learn the cowardice and weakness of our hearts.

The Agony in the Garden (14:32-42)

Darkness had settled over the land. It was Thursday night running into Friday morning. When they came to an enclosed piece of ground called Gethsemane, the Lord Jesus left eight of the disciples near the entrance. He took Peter, James, and John a little further with Him into the garden. There He experienced an overpowering burden of distress and horror. It was doubtless the unspeakable revulsion of His holy soul as He anticipated becoming a sin-offering for us. We cannot conceive what it meant to Him, the Sinless One, to be made sin for us.

He left the three disciples with instructions to stop there and stay awake. He moved on further into the garden—alone. Thus would He go to the cross alone, bearing the awful judgment of God against our sins.

With wonder and amazement, we see the Lord Jesus prostrate on the ground, praying to God. And what was the burden of His prayer? Was He asking to be excused from going to the cross? Not at all: this was the purpose of His coming into the world.

First, He prayed that if it were possible, the hour might pass away from Him. In other words, if there was any other way by which sinners could be saved than by His death, burial and resurrection, let God make that way manifest. The heavens were silent. There was no answer. From this we learn that there was no other way in which we could be redeemed.

Again, He prayed, "Abba, Father, all things are possible for you. Remove this cup from me. Yet not what I will, but what you will." Notice that He addressed God as His beloved Father with whom all things are possible. Here it was not so much a matter of physical possibility as of moral. Could the Almighty Father find any other righteous basis upon which He could save ungodly sinners? The silent heavens indicated that there was no other way. The Holy Son of God must bleed that sinners might from sin be freed!

Returning to the three disciples, He found them asleep. It is a sad commentary on fallen human nature. Jesus turned and warned Peter against sleeping in that crucial hour. Only recently Peter had boasted of his undying steadfastness. Now He couldn't even stay awake. If a person cannot pray for one hour, it is unlikely that they will be able to resist temptation in the moment of extreme pressure. No matter how enthusiastic their spirit may be, they must reckon with the frailty of their flesh.

Three times the Lord Jesus returned to find the disciples asleep. Then He said, "Are you still sleeping and resting? It is enough! The hour has come; behold, the Son of Man is being betrayed into the hands of sinners." (Phillips).

With that, they got up as if to go forth. But they didn't have to go far.

The Betrayal (14:43-52)

Judas had already entered the garden with a posse. His cohorts were carrying swords and clubs, as if they were going to capture a dangerous felon.

Judas had prearranged a sign. He would kiss the one whom they should seize. So he strode up to Jesus, addressed Him as Rabbi, and kissed Him effusively.

Two different words for "kiss" are used in the original. In verse 44 it is the kiss of respect or affection. In the next verse, it is the lover's kiss, gushing and demonstrative.

Why did Judas betray the Lord? Was he disappointed that Jesus had not seized the reigns of government? Were his hopes dashed for a place of prominence in the kingdom? Was he overcome by greed? All of these might have contributed to his infamous deed.

Peter was using carnal weapons to fight a spiritual warfare.

The armed henchmen of the betrayer stepped forward and arrested the Lord. Peter quickly took his sword and with a deft stroke, sliced off the ear of the servant of the high priest. It was a natural reaction, not a spiritual one. Peter was using carnal weapons to fight a spiritual warfare. The Lord rebuked Peter and miraculously restored the ear, as we read in Luke 22:51. Jesus then reminded His captors how incongruous it was for them to take Him by force! He had been with them in the temple teaching. Why hadn't they seized Him then? He knew the answer. The Scriptures must be fulfilled which prophesied that He would be betrayed (Psalm 41:9), arrested (Isaiah 53:7), manhandled (Psalm 22:12) and forsaken (Zechariah 13:7).

Mark is the only evangelist who gives us the incident recorded in verses 51 and 52. Possibly he himself was the young man who, in his frenzy to escape, left his linen covering in the grasp of the armed men. The linen cloth was not a regular garment but a piece of cloth that he had picked up quickly and used as an improvised covering.

"Probably this picturesque incident is added to show how completely Jesus was forsaken in the hours of His peril and pain. He surely knew what it was to suffer alone" (Erdman).

Jesus Before the High Priest (14:53-54)

The record of the ecclesiastical trial extends from verse 53 to 15:1 and is divided into three parts:

1. Trial before the high priest (vv. 53-54)
2. Midnight meeting of the Sanhedrin (vv. 55-65)
3. Meeting of the Sanhedrin in the morning (15:1)

It is generally agreed that Mark records the trial before Caiaphas in v. 53. The trial before Annas is found in John 18:13, 19-24.

Peter trailed the Lord Jesus to the court of the high priest; following at what he thought would be a safe distance. Someone has outlined his downfall as follows:

- He first fought—misdirected enthusiasm
- He then fled—cowardly withdrawal
- Finally he followed afar off—half-hearted discipleship by night

He sat by the fire with the officers, warming himself with the enemies of his Lord.

Jesus Before the Sanhedrin (14:55-65)

Although it is not clearly stated, verse 55 seems to begin the account of a midnight meeting of the Sanhedrin. The body of seventy-one religious leaders was presided over by the high priest.

On this particular night, the Pharisees, Sadducees, scribes, and elders who comprised the Sanhedrin showed an utter disregard for the rules under which they operated.

- They were not supposed to meet at night or during any of the Jewish feasts.
- They were not supposed to bribe witnesses to commit perjury.
- A death verdict was not to be carried out until a night had elapsed.
- Unless they met in the Hall of Hewn Stone, in the temple area, their verdicts were not binding.

"In their eagerness to eliminate Jesus, the Jewish authorities did not hesitate to stoop to breaking their own laws" (Selected).

The determined efforts of the religious leaders produced a group of false witnesses but they failed to produce united testimony. Some misquoted the Lord as threatening to destroy the temple made with hands and to rebuild another in three days, made without hands. What Jesus actually said is found in John 2:19. They purposely confused the temple in Jerusalem with the temple of His Body.

When the high priest first questioned Him, Jesus did not reply. But when he asked the Lord under oath (Matthew 26:63) whether He was the Messiah, the Son of the Blessed, the Savior replied that He was. The Lord thus acted in obedience to Leviticus 5:1. Then, as if to remove any possible doubt as to who He claimed to be, the Lord Jesus told the high priest point blank that he would yet see the Son of Man sitting at the right hand of Power and coming back to earth with the clouds of heaven. What did He mean? He meant that the high priest would yet see Him openly manifested as God. During His first advent, the glory of His deity was veiled in a human body. But when He comes again in power and great glory, the veil will be removed and everyone will know then exactly who He is.

The high priest understood what Jesus meant. He tore his garment as a sign of his righteous indignation against this blasphemy (as he thought). The one Israelite who should have been most ready to recognize and receive the Messiah was loudest in his condemnation. But not he alone; all the members of the Sanhedrin agreed that Jesus had spoken blasphemy, and they judged Him worthy of death.

The scene that followed was grotesque in the extreme. Some members of the Sanhedrin spat upon the Son of God, blindfolded Him, then challenged

Him to name His assailants. It is almost incredible that the Worthy Savior should have to endure such contradiction of sinners against Himself. The officers (temple police) joined in the scandal by slapping Him (or by beating Him with rods, as in some versions).

Peter's Denials (14:66-72)

Peter was waiting below in the inner court of the building. One of the high priest's serving maids happened to walk by. She peered intently at him, then charged him with being a follower of the Nazarene, Jesus. The pathetic disciple pretended complete ignorance of her charge, then moved to the porch in time to hear a cock crow. It was a ghastly moment. Sin was taking its terrible toll.

The maid saw him again and pointed him out as a disciple of Jesus. Peter made another cold denial, and probably wondered why people didn't leave him alone.

Then the crowd said to Peter, "Certainly you are one of them, for you are a Galilean." With cursing and swearing, Peter defiantly stated that he didn't know the Man. No sooner were the words out of his mouth than the cock crew. The world of nature seemed thus to protest the cowardly lie.

In a flash Peter realized that the Lord's prediction had come to pass. He broke down and wept bitterly.

It is significant that all four Gospels record the denials of Peter. We must all learn the lesson that self confidence leads to humiliation. We must learn to distrust self and to lean completely on the power of God to sustain us.

LESSON 11 EXAM

Use the exam sheet at the back of the course to complete your exam.

1. **The Lord Jesus was crucified at the time of Passover because**
   A. that was when the religious leaders wanted to arrest, try and execute Him.
   B. that was the time of a national religious holiday and the Roman governor wanted to make a public example of Jesus.
   C. God overruled human plans to fulfill His own purposes as to the time of Christ's death.
   D. the time of Passover was a relatively safe time to plan the crucifixion

2. **Mark's purpose in inserting the story of the anointing of Jesus by the woman seems to emphasize**
   A. the value Jesus placed on the woman's act of devotion.
   B. the contrast between the woman's evaluation of Jesus and the evaluation placed on Him by Judas.
   C. the importance of caring for the poor.
   D. the carefulness of Mark in following a strict chronological order.

3. **According to Mark's account, Judas agreed to betray Jesus**
   A. for the honor Judas would receive.
   B. for money.
   C. in exchange for a piece of property near Jerusalem known as Aceldama.
   D. for three hundred pence (a year's wages).

4. **When Jesus announced that one of the Twelve would betray Him**
   A. they all denied such a thing to be possible.
   B. Peter said that though all the others betrayed the Lord he never would.
   C. each one asked if he would be the one.
   D. Philip said it would be better for such a man never to have been born.

5. **In the Lord's Supper, the bread signifies**
  A. the Lord's body given for us on the Cross.
  B. the various backgrounds of those who feast on Christ.
  C. the lack of festive joy until the Lord returns to earth.
  D. the ultimate sharing of all the redeemed in the marriage supper of the Lamb.

6. **The Gethsemane agony of the Lord Jesus was**
  A. witnessed by all the disciples, except Judas.
  B. an expression of the Lord's distress at the thought of being made sin for us.
  C. brought on by physical exhaustion.
  D. disturbing to Peter, James and John.

7. **In His Gethsemane prayer the Lord Jesus addressed God as**
  A. Abba, Father.
  B. Elohim ("My God, My God".)
  C. Jehovah Jireh.
  D. Almighty God.

8. **The author mentions several illegal proceedings of the Sanhedrin's trial of Jesus. One of them was**
  A. having the high priest preside.
  B. not giving Jesus a chance to defend Himself.
  C. making Jesus answer under oath.
  D. holding the trial at night.

9. **When before the Sanhedrin the Lord Jesus**
  A. refused to answer any questions even when put under oath.
  B. told the High Priest he was "a whited wall," that is, a hypocrite.
  C. confessed that He was Christ.
  D. again promised to rebuild the Temple in three days if it were destroyed.

10. **Peter's denial of the Lord**
  A. was premeditated on his part.
  B. is recorded only in Mark and John.
  C. should teach us complete distrust in self.
  D. was brought about by the deliberate efforts of the authorities to ruin his testimony.

What Do You Say?

Why wasn't the value of the perfume wasted when the woman used it to anoint Jesus?

More Than Conqueror

Mark 15, 16

The Sanhedrin Meets (15:1)

The first verse describes a morning meeting of the Sanhedrin, perhaps convened to validate the illegal action of the night before. As a result, Jesus was bound and taken to Pilate, the Roman Governor of Palestine.

Jesus Before Pilate (15:2-5)

Up to now, Jesus had been on trial before the religious leaders on a charge of blasphemy. Now He was taken before the civil court on a charge of treason. The civil trial took place in three stages:

1. Before Pilate
2. Before Herod
3. Before Pilate again

Pilate asked the Lord Jesus if He were the King of the Jews. If He were, this presumably meant that He was dedicated to the overthrow of Caesar; and was thus guilty of treason. The Lord's answer, "You have said so," appears to put the words back in Pilate's mouth, but they were not a denial.

The chief priests poured out a torrent of charges against Jesus. Pilate couldn't get over His poise in the face of such overwhelming accusations. He asked Him why He didn't defend Himself, but Jesus refused to answer His critics.

Barabbas Chosen over Jesus (15:6-15)

It was the custom for the Roman Governor to release some Jewish prisoner at this feast time—sort of a political gesture to the unhappy people. One such eligible prisoner was Barabbas, guilty of insurrection and murder. When Pilate offered to release Jesus, taunting the chief priests, the people were primed to ask for Barabbas. The very ones who were charging Jesus with treason against Caesar were asking the release of a man who was actually guilty of that crime. The position of the chief priests was irrational and ludicrous—but sin is like that.

Pilate naively asked what he should do with the One whom they called King of the Jews. The people chanted savagely, "Crucify Him!" Pilate demanded a reason, but there was no reason. Mob hysteria was rising. All they would shout was, "Crucify Him!"

And so the spineless Pilate did what they wanted—he released Barabbas, scourged Jesus and delivered Him over to the soldiers for crucifixion. It was an enormous miscarriage of justice. It was a monstrous verdict of unrighteousness. And yet it was a parable of our redemption—the guiltless One delivered to die in order that the guilty might go free.

Mockery of the Soldiers (15:16-21)

The soldiers marched Jesus off to the courtyard of the Governor's residence. After assembling their entire military unit, they staged a mock coronation for the King of the Jews.

If they had only known! It was God the Son they clothed with purple. It was their own Creator they crowned with thorns. It was the Sustainer of the universe they mocked as King of the Jews. It was the Lord of life and glory they smote on the head. They spat upon the Prince of peace. They mockingly bowed their knees to the King of kings and Lord of lords.

When their crude jests were over, they put His own clothes back on Him, then took Him out to crucify Him. Mark mentions here that the soldiers ordered a passerby, Simon of Cyrene, to carry the cross.

Who was this Simon? He came from a city in the north of Africa. He may have been black but was more probably a Hellenistic Jewish man. He had two sons, Alexander and Rufus, who were probably believers (if Rufus is the same one mentioned in Romans 16:13). In bearing the cross after Jesus, he gave us a picture of what should characterize us as disciples of the Savior.

The Crucifixion (15:22-32)

In describing the crucifixion, the Spirit of God does not dwell on the extreme cruelty of this mode of execution, or of the terrible suffering it entailed. With remarkable restraint, He describes it simply and unemotionally.

The exact location is unknown today. The traditional site is at the Church of the Holy Sepulcher. Though this site is inside the walls of the city, its advocates contend that it was outside the walls at the time of Christ.

Another supposed site is Gordon's Calvary, north of the city walls and adjoining a garden area.

Golgotha is the Aramaic name meaning "skull." Calvary is the Greek name. Perhaps the area was shaped like a skull. Perhaps it received the name because it was a place of execution.

The guiltless One died in order that the guilty might go free.

The soldiers offered Jesus wine mixed with myrrh. This would have acted as a drug, dulling His senses. He would not take it. He was determined to bear man's sins in His full consciousness.

The soldiers gambled for the clothes of those who were crucified. When they took the clothes of the Savior, they took just about everything material that He owned.

It was 9:00 a.m. when they crucified Him. Over His head they had put the title "The King of the Jews." (Mark does not give the full superscription but contents himself with the substance of it, see Matthew 27:37; Luke 23:38; John 19:19.) Two robbers were crucified with Him, one on each side—just as Isaiah had predicted that He would be associated with criminals in His death (Isaiah 53:12).

The Lord Jesus was mocked by:

- The passers-by (vv. 29-30)
- The chief priests and scribes (vv. 31-32)
- The two robbers (v. 32)

The passers-by were probably Jewish who were ready to keep the Passover inside the city itself. Outside they paused long enough to hurl an insult at the Paschal Lamb. They misquoted Him as threatening to destroy their beloved temple and to rebuild it in three days. If He was so great, let Him save Himself by coming down from the cross.

The chief priests and scribes scorned His claim to save others. He couldn't even save Himself. "He saved others; He cannot save Himself." It was viciously cruel, yet unintentionally true.

> Himself He could not save,
> He on the cross must die
> Or mercy could not come
> To ruined sinners nigh.
> Yes, Christ the Son of God must bleed
> That sinners might from sin be freed.

It was true in the Lord's life and it is true in ours too. We can't save others while seeking to save ourselves.

The religious leaders also challenged Him to come down from the cross if He were the Messiah, the King of Israel. Then they would believe, they said. Let us see and we will believe. But God's order is, "Believe and then you will see." Even the criminals reproached Him!

> Why? What hath my Lord done?
> What makes this rage and spite?
> He made the lame to run,
> He gave the blind their sight.
> Sweet injuries!
> Yet they at these
> Themselves displease,
> And 'gainst Him rise.

The Three Hours of Darkness (15:33-41)

Between noon and three o'clock the whole land was shrouded in darkness. Jesus was then bearing the full judgment of God against our sins. He suffered spiritual desolation and separation from God. No mortal mind will ever be able to understand the agony He endured when His soul was made a sacrifice for sin.

At the close of His agony, Jesus cried out in Aramaic with a loud voice, "My God, my God, why have you forsaken me?" God had forsaken Him because in His holiness He must dissociate Himself from sin. The Lord Jesus had identified Himself with our sins and was paying the penalty in full.

Some of the cruel mob suggested He was calling for Elijah when He said, "Eloi, Eloi." It was just a jest on their part. What a time to be jesting! As a final indignity, one of them soaked a sponge in sour wine and offered it to Him on the end of a pole.

Jesus spoke out with strength and triumph, then dismissed His spirit. His death was an act of His own will, not an involuntary collapse.

> **Jesus spoke out with strength and triumph.**

At that moment, the veil of the temple was torn form the top to the bottom. This was an act of God indicating that by the death of Christ, access into the sanctuary of God was henceforth the privilege of all believers (see Hebrews 10:19-22). A great new era had been ushered in. It would be an era of nearness to God, not of distance from Him.

The Gentile centurion's confession, while noble, did not necessarily acknowledge Jesus as equal with God. The Roman officer recognized Him as a Son of God. No doubt he had a sense of history being made. But whether his faith was genuine is not clear.

Mark mentions that certain women remained at the cross. It must be confessed that the women shine brightly in the Gospel narratives. Considerations of personal safety drove the men into hiding. The devotion of the women put love to Christ above their own welfare.

The Burial (15:42-47)

The Sabbath began at sunset on Friday night. The day before a Sabbath or other festival was known as the Preparation.

The necessity for prompt action probably emboldened Joseph of Arimathaea to ask Pilate for permission to bury the body of Jesus. Joseph was a devout Jewish man, perhaps a member of the Sanhedrin (Luke 23:51). (See also Matthew 27:57; Mark 15:43; Luke 23:50; John 19:38.)

Pilate could hardly believe that Jesus was dead already. When the centurion confirmed the fact, the Governor gave the corpse to Joseph.

Two different words are used for the body of Jesus in this section. Joseph asked for the *body* of the Lord Jesus (v. 43). Pilate granted the *corpse* to him (v. 45).

With loving care, Joseph (and Nicodemus, John 19:38-39) embalmed the body, wrapped it in linen, and then put it in a new tomb belonging to himself. The tomb was a small room carved out of solid rock. The door was sealed with a coin-shaped stone that probably moved in a groove carved out of stone.

In life, no house, no home
My Lord on earth might have;
In death, no friendly tomb
But what a stranger gave.
What may I say?
Heaven was His home;
But mine the tomb
Wherein He lay.

Again the women, that is, the two Marys, are mentioned as being present. We admire them for their unflagging and fearless affection.

We are told that the majority of missionaries today are women. Where are the men?

Women at the Grave (16:1-8)

On Saturday evening the two Marys bought spices with which they could embalm the body of Jesus. They knew it would not be easy. They knew a great stone had been rolled across the mouth of the tomb. They knew about the Roman seal and the guard of soldiers. But love leaps over mountains of difficulties to reach the object of its affection.

Early on Sunday morning, they were wondering out loud who would roll away the stone from the sepulcher. They looked up and saw that it had already been done! How often it happens when we are intent on honoring the Savior that difficulties are removed before we get to them.

Inside the tomb they saw an angelic being with the appearance of a young man in white. He quickly dispelled their fears with the announcement of the resurrection. Jesus had risen. The tomb was empty.

The angel then commissioned them as heralds of the resurrection. They were to tell the disciples and Peter that Jesus would meet them in Galilee.

Notice the words "and Peter." The disciple who had denied His Lord was singled out for special mention. The risen Redeemer had not disowned him but still loved him and longed to see him again. A special work of restoration needed to be done. The wandering sheep must be brought back into fellowship with the Shepherd. The backslider must return to the Father's house.

The women fled from the tomb with mingled shock and panic. They were too afraid to tell anyone what had happened. This is not surprising. The wonder is that they had been so brave and loyal and devoted up to now.

Appearance to Mary Magdalene (16:9-11)

(Note: Some of the oldest manuscripts of the Bible omit the rest of the chapter.)

The Savior's first appearance was to Mary Magdalene. The first time she had met Jesus, He had cast seven demons out of her. From then on she served Him lovingly with her possessions. She witnessed the crucifixion, and saw where His body was laid.

Jesus had risen. The tomb was empty.

From the other Gospels, we learn that after finding the tomb empty on Sunday morning, she ran and told Peter and John. They came back with her to the sepulcher to find it empty, as she had told them. They returned to their home but she stayed at the empty tomb. It was then that Jesus appeared to her.

Again she ran back to the city to share the good news with the sorrowing disciples. For them it was too good to be true. They wouldn't believe it.

Appearance to Two on Way to Emmaus (16:12-13)

The full account of this appearance is found in Luke 24:13-31. Here we read that He was manifested in another form to the two disciples on the Road to Emmaus. To Mary He had appeared as a gardener. Now He seemed like a fellow-traveler. But it was the same Jesus in His glorified body.

When the two disciples returned to Jerusalem and reported their fellowship with the risen Savior, they met the same disbelief that Mary had encountered.

Appearance to the Eleven (16:14-18)

This appearance to the eleven took place that same Sunday evening (1 Corinthians 15:5; Luke 24:36; John 20:19-24). Although the disciples are referred to as "the eleven," only ten were present. Thomas was absent on this occasion. Jesus rebuked His own for their refusal to accept the reports of His resurrection from Mary and the others.

Verse 15 records the commission that was given by the Lord on the eve of His ascension. There is thus an interval between verses 14 and 15.

The disciples were commanded to preach the Gospel to the whole creation. The Savior's goal was world evangelization. He purposed to accomplish it with eleven disciples who would literally forsake all to follow Him.

There would be two results of the preaching. Some would believe, be baptized and be saved; some would disbelieve and be condemned.

Verse 16 is understood by some to teach the necessity of baptism for salvation. We know it cannot mean that for the following reasons:

- The thief on the cross was not baptized; yet he was assured of being in Paradise with Christ (Luke 23:43).
- The Gentiles in Caesarea were baptized after they were saved (Acts 10:44-48).
- Jesus Himself did not baptize (John 4:1-2)—a strange omission if baptism were necessary for salvation.
- Paul thanked God that he baptized very few of the Corinthians (1 Corinthians 1:14-16)—a strange thing to be thankful for if baptism were essential for salvation.
- Approximately 150 passages in the New Testament state that salvation is by faith alone. No verse or few verses could contradict this overwhelming testimony.
- Baptism is connected with death and burial in the New Testament, not with spiritual birth.

What then does verse 16 mean? We believe it mentions baptism as the expected outward expression of belief. If a person believed, they were expected to be baptized. Baptism is not a condition of salvation, but an outward proclamation that the person has been saved.

In verses 17 and 18, Jesus describes certain miracles that would accompany those who believe the Gospel. As we read the verses, the obvious question is, "Do these signs exist today?"

We believe that these signs were intended primarily for the apostolic age, before the complete Bible was available in written form. Most of these signs are found in the book of Acts:

- Demons cast out (Acts 8:7; 16:18; 19:11-16)
- New tongues (Acts 2:4-11; 10:46; 19:6)
- Handle serpents (Acts 28:5)
- Drink poison without harmful effects—not recorded in Acts but attributed to John and Barnabas by Eusebius (a Roman historian)
- Lay hands on the sick for healing (Acts 3:7; 19:11; 28:8-9)

What was the purpose of these miracles? We believe the answer is found in Hebrews 2:3-4. Before the New Testament was available in completed form, people would ask the apostles and others for proof that the Gospel was divine. To confirm the preaching, God bore witness with signs and wonders and various gifts of the Holy Spirit.

The need for these miraculous signs is gone today. We have the complete Bible. If people won't believe that, they wouldn't believe anyway.

Mark did not say that the miracles would continue. The words "to the end of the age" are not found here as they are in Matthew 28:18-20.

Ascension (16:19-20)

Forty days after His resurrection, our Lord Jesus ascended to heaven and sat down at the right hand of God. This is the place of honor and of power.

In obedience to His command, the disciples went forth like flaming fires, preaching the Gospel and winning people to the Savior. The power of God was with them. The promised miracles accompanied their preaching, confirming the Word they spoke.

Here the narrative ends—with Christ in heaven, with a few committed disciples on earth burdened for world evangelization and giving themselves entirely to it, and with results of eternal consequences.

We are entrusted with the great commission in our generation. Our task is to reach every creature with the Gospel. One half of all the people who have ever lived are living today. As the population explodes, the task increases. But the method is always the same—devoted disciples with unlimited love for Christ who count no sacrifice too great for Him.

The will of God is the evangelization of the world. What are you doing about it?

LESSON 12 EXAM

Use the exam sheet at the back of the course to complete your exam.

1. **The Lord Jesus was arraigned before Pilate on a charge of**
   A. blasphemy.
   B. treason.
   C. insurrection and riot.
   D. heresy.

2. **The man that the mob chose rather than Jesus was**
   A. a Hellenistic Jew.
   B. an insurrectionist and a murderer.
   C. a thief and a robber.
   D. chief of the brigands.

3. **Jesus was crucified at Golgotha,**
   A. "the place of a skull."
   B. "the field of blood."
   C. "the mount of the Lord."
   D. "the hill of shame."

4. **Those who said, mockingly (but truly) of Christ that "He saved others, Himself He cannot save"**
   A. were the chief priests and the scribes, the leaders of religion.
   B. the Greeks, representatives of world culture and progress.
   C. the Roman authorities, the advocates of politics and government.
   D. the common people, representing the masses.

5. **The death of the Lord Jesus**
   A. resulted from involuntary collapse.
   B. was an act of His own will.
   C. was brought about by profuse bleeding from His wounds.
   D. resulted from the spear of the Roman soldier which pierced His heart.

6. **According to Matthew 27:57, Mark 15:43, and Luke 23:50, Joseph of Arimathea was**
   A. rich, according to Matthew.
   B. a prominent council member, according to Mark.
   C. a good and upright man, according to Luke.
   D. All of the above.

7. **According to Mark 16:1, the two Marys who came to the tomb on the resurrection morning were**
   A. Mary Magdalene and Mary the mother of James.
   B. Mary the mother of Jesus and Mary of Bethany.
   C. Mary the wife of Clopas and Mary the sister of Lazarus.
   D. Mary Magdalene and Mary of Bethany.

8. **The first resurrection appearance of the Lord Jesus was to**
   A. the disciples in the upper room.
   B. Simon Peter.
   C. Mary Magdalene.
   D. Peter and John at the tomb.

9. **When the Lord Jesus first appeared in the upper room**
   A. all the disciples were present.
   B. all were present except Judas and Thomas.
   C. Simon Peter was absent.
   D. some one hundred and twenty believers were present.

10. **The statement "he who believes and is baptized will be saved"**
    A. makes baptism essential to salvation.
    B. shows that baptism is the expected outward expression of inward belief.
    C. contradicts the overwhelming testimony of the New Testament which state that salvation is by faith alone.
    D. proves Paul to be a poor evangelist since he usually did not personally baptize his converts.

What Do You Say?

What is the significance of the veil of the temple being torn from the top to the bottom when Jesus died?

_____

_____

_____

_____

For even the Son of Man came not to be served but to serve, and to give his life as a ransom for many.

—Mark 10:45

EXAM SHEET
The Gospel of Mark

DIRECTIONS: 1) Use a pen or pencil. 2) Completely fill in the circle.
3) Erase or X any answer you wish to change. 4) Try not to make stray marks.

INCORRECT: Ⓐ̶ Ⓐ̌ Ⓒ Ⓓ̶
CORRECT: Ⓐ ● Ⓒ Ⓓ

1-1. Ⓐ Ⓑ Ⓒ Ⓓ	4-1. Ⓐ Ⓑ Ⓒ Ⓓ	7-1. Ⓐ Ⓑ Ⓒ Ⓓ	10-1. Ⓐ Ⓑ Ⓒ Ⓓ
2. Ⓐ Ⓑ Ⓒ Ⓓ	2. Ⓐ Ⓑ Ⓒ Ⓓ	2. Ⓐ Ⓑ Ⓒ Ⓓ	2. Ⓐ Ⓑ Ⓒ Ⓓ
3. Ⓐ Ⓑ Ⓒ Ⓓ	3. Ⓐ Ⓑ Ⓒ Ⓓ	3. Ⓐ Ⓑ Ⓒ Ⓓ	3. Ⓐ Ⓑ Ⓒ Ⓓ
4. Ⓐ Ⓑ Ⓒ Ⓓ	4. Ⓐ Ⓑ Ⓒ Ⓓ	4. Ⓐ Ⓑ Ⓒ Ⓓ	4. Ⓐ Ⓑ Ⓒ Ⓓ
5. Ⓐ Ⓑ Ⓒ Ⓓ	5. Ⓐ Ⓑ Ⓒ Ⓓ	5. Ⓐ Ⓑ Ⓒ Ⓓ	5. Ⓐ Ⓑ Ⓒ Ⓓ
6. Ⓐ Ⓑ Ⓒ Ⓓ	6. Ⓐ Ⓑ Ⓒ Ⓓ	6. Ⓐ Ⓑ Ⓒ Ⓓ	6. Ⓐ Ⓑ Ⓒ Ⓓ
7. Ⓐ Ⓑ Ⓒ Ⓓ	7. Ⓐ Ⓑ Ⓒ Ⓓ	7. Ⓐ Ⓑ Ⓒ Ⓓ	7. Ⓐ Ⓑ Ⓒ Ⓓ
8. Ⓐ Ⓑ Ⓒ Ⓓ	8. Ⓐ Ⓑ Ⓒ Ⓓ	8. Ⓐ Ⓑ Ⓒ Ⓓ	8. Ⓐ Ⓑ Ⓒ Ⓓ
9. Ⓐ Ⓑ Ⓒ Ⓓ	9. Ⓐ Ⓑ Ⓒ Ⓓ	9. Ⓐ Ⓑ Ⓒ Ⓓ	9. Ⓐ Ⓑ Ⓒ Ⓓ
10. Ⓐ Ⓑ Ⓒ Ⓓ	10. Ⓐ Ⓑ Ⓒ Ⓓ	10. Ⓐ Ⓑ Ⓒ Ⓓ	10. Ⓐ Ⓑ Ⓒ Ⓓ
2-1. Ⓐ Ⓑ Ⓒ Ⓓ	5-1. Ⓐ Ⓑ Ⓒ Ⓓ	8-1. Ⓐ Ⓑ Ⓒ Ⓓ	11-1. Ⓐ Ⓑ Ⓒ Ⓓ
2. Ⓐ Ⓑ Ⓒ Ⓓ	2. Ⓐ Ⓑ Ⓒ Ⓓ	2. Ⓐ Ⓑ Ⓒ Ⓓ	2. Ⓐ Ⓑ Ⓒ Ⓓ
3. Ⓐ Ⓑ Ⓒ Ⓓ	3. Ⓐ Ⓑ Ⓒ Ⓓ	3. Ⓐ Ⓑ Ⓒ Ⓓ	3. Ⓐ Ⓑ Ⓒ Ⓓ
4. Ⓐ Ⓑ Ⓒ Ⓓ	4. Ⓐ Ⓑ Ⓒ Ⓓ	4. Ⓐ Ⓑ Ⓒ Ⓓ	4. Ⓐ Ⓑ Ⓒ Ⓓ
5. Ⓐ Ⓑ Ⓒ Ⓓ	5. Ⓐ Ⓑ Ⓒ Ⓓ	5. Ⓐ Ⓑ Ⓒ Ⓓ	5. Ⓐ Ⓑ Ⓒ Ⓓ
6. Ⓐ Ⓑ Ⓒ Ⓓ	6. Ⓐ Ⓑ Ⓒ Ⓓ	6. Ⓐ Ⓑ Ⓒ Ⓓ	6. Ⓐ Ⓑ Ⓒ Ⓓ
7. Ⓐ Ⓑ Ⓒ Ⓓ	7. Ⓐ Ⓑ Ⓒ Ⓓ	7. Ⓐ Ⓑ Ⓒ Ⓓ	7. Ⓐ Ⓑ Ⓒ Ⓓ
8. Ⓐ Ⓑ Ⓒ Ⓓ	8. Ⓐ Ⓑ Ⓒ Ⓓ	8. Ⓐ Ⓑ Ⓒ Ⓓ	8. Ⓐ Ⓑ Ⓒ Ⓓ
9. Ⓐ Ⓑ Ⓒ Ⓓ	9. Ⓐ Ⓑ Ⓒ Ⓓ	9. Ⓐ Ⓑ Ⓒ Ⓓ	9. Ⓐ Ⓑ Ⓒ Ⓓ
10. Ⓐ Ⓑ Ⓒ Ⓓ	10. Ⓐ Ⓑ Ⓒ Ⓓ	10. Ⓐ Ⓑ Ⓒ Ⓓ	10. Ⓐ Ⓑ Ⓒ Ⓓ
3-1. Ⓐ Ⓑ Ⓒ Ⓓ	6-1. Ⓐ Ⓑ Ⓒ Ⓓ	9-1. Ⓐ Ⓑ Ⓒ Ⓓ	12-1. Ⓐ Ⓑ Ⓒ Ⓓ
2. Ⓐ Ⓑ Ⓒ Ⓓ	2. Ⓐ Ⓑ Ⓒ Ⓓ	2. Ⓐ Ⓑ Ⓒ Ⓓ	2. Ⓐ Ⓑ Ⓒ Ⓓ
3. Ⓐ Ⓑ Ⓒ Ⓓ	3. Ⓐ Ⓑ Ⓒ Ⓓ	3. Ⓐ Ⓑ Ⓒ Ⓓ	3. Ⓐ Ⓑ Ⓒ Ⓓ
4. Ⓐ Ⓑ Ⓒ Ⓓ	4. Ⓐ Ⓑ Ⓒ Ⓓ	4. Ⓐ Ⓑ Ⓒ Ⓓ	4. Ⓐ Ⓑ Ⓒ Ⓓ
5. Ⓐ Ⓑ Ⓒ Ⓓ	5. Ⓐ Ⓑ Ⓒ Ⓓ	5. Ⓐ Ⓑ Ⓒ Ⓓ	5. Ⓐ Ⓑ Ⓒ Ⓓ
6. Ⓐ Ⓑ Ⓒ Ⓓ	6. Ⓐ Ⓑ Ⓒ Ⓓ	6. Ⓐ Ⓑ Ⓒ Ⓓ	6. Ⓐ Ⓑ Ⓒ Ⓓ
7. Ⓐ Ⓑ Ⓒ Ⓓ	7. Ⓐ Ⓑ Ⓒ Ⓓ	7. Ⓐ Ⓑ Ⓒ Ⓓ	7. Ⓐ Ⓑ Ⓒ Ⓓ
8. Ⓐ Ⓑ Ⓒ Ⓓ	8. Ⓐ Ⓑ Ⓒ Ⓓ	8. Ⓐ Ⓑ Ⓒ Ⓓ	8. Ⓐ Ⓑ Ⓒ Ⓓ
9. Ⓐ Ⓑ Ⓒ Ⓓ	9. Ⓐ Ⓑ Ⓒ Ⓓ	9. Ⓐ Ⓑ Ⓒ Ⓓ	9. Ⓐ Ⓑ Ⓒ Ⓓ
10. Ⓐ Ⓑ Ⓒ Ⓓ	10. Ⓐ Ⓑ Ⓒ Ⓓ	10. Ⓐ Ⓑ Ⓒ Ⓓ	10. Ⓐ Ⓑ Ⓒ Ⓓ

Form Identifier — DO NOT MARK

ER11MK-AK20

Write It Out!
The Gospel of Mark

These questions will be reviewed and responded to by an Emmaus Connector.

1. HEAD: Mark presents the Lord Jesus Christ as a Servant. Give specific instances you saw this to be true in the book of Mark.

2. HEART: How does this course affect your perspective of or feelings towards God, yourself, or others?

3. HANDS: How does this course affect your actions today?

PRAYER REQUESTS OR QUESTIONS?

Your Information

First Name:_____ Last Name:_____

Address:_____

City:_____ State:_____ ZIP:_____

Email Address:_____

Emmaus Connector: (If known) _____

Institution: (If applicable) _____

Cell Location: (If applicable) _____ ID # (If applicable) _____

By submitting this exam sheet, you agree to allow Emmaus Worldwide and/or your Emmaus Connector to keep your information securely on file in order to mail back your exam sheet, produce your completion certificates, and keep your transcript up-to-date.